DARWIN'S BRIDGE

DARWIN'S BRIDGE

Uniting the Humanities and Sciences

Edited by

Joseph Carroll,
Dan P. McAdams,
and
Edward O. Wilson

OXFORD
UNIVERSITY PRESS

OXFORD
UNIVERSITY PRESS

Oxford University Press is a department of the University of Oxford. It furthers the University's objective of excellence in research, scholarship, and education by publishing worldwide. Oxford is a registered trade mark of Oxford University Press in the UK and certain other countries.

Published in the United States of America by Oxford University Press
198 Madison Avenue, New York, NY 10016, United States of America.

Library of Congress Cataloging-in-Publication Data
Names: Carroll, Joseph, 1949- editor.
Title: Darwin's bridge : uniting the humanities and sciences / edited by Joseph Carroll, Dan P. McAdams, and Edward O. Wilson.
Description: New York : Oxford University Press, 2016.
Identifiers: LCCN 2015041641 | ISBN 9780190231217
Subjects: LCSH: Social ecology. | Science and the humanities. | Darwin, Charles, 1809–1882—Influence.
Classification: LCC HM861 .D37 2016 | DDC 304.2—dc23
LC record available at http://lccn.loc.gov/2015041641

9 8 7 6 5 4 3 2 1

Printed by Sheridan Books, Inc., United States of America

For Henry Harpending, 1944–2016

CONTENTS

PART THREE: *Ancient Markings*

PART FOUR: *Integrative Psychology*

PART FIVE: *A Biocultural Perspective on Literature*

FOREWORD

Alice Dreger

Perhaps because my mate and I are both scholars, our son didn't leave the "why?" stage according to the typical childhood developmental schedule. By the age of four, if we answered his latest "why" question with "I don't know," rather than give up, he would push harder: "Well, do you have any *guesses*?"

When he was five years old, we took him in the car from Michigan to New York to visit my family. Around the 12th hour, just as we were stuck in New Jersey traffic heading onto the George Washington Bridge, he asked "why?" about something at a highway construction site. Exhausted from the trip and genuinely unsure about why the construction site worked as it did, I decided to preempt the inevitable follow-up question, curtly answering him, "Honey, I don't know *why* and I don't have any *guesses*, either."

A short moment of quiet ensued. Just as I thought I had managed liberation from this latest line of questioning, a small voice popped up from the back seat: "Well, then, do you have any *suspicions*?"

I often think of this amusing interchange when I'm talking to a journalist about something I've found, and the journalist is growing increasingly impatient with my tendency to stick to a historical description that lacks deeper causal explanation. *But why?*, reporters will ask. *Why did that person do that? Why did his or her colleagues react as they did? Why is our culture the way it is? Why has it changed?*

It seems impossible to get them to accept the answer, "Sometimes, as historians, all we can do is describe *what* happened." They don't want to hear that motivation for human action and the reasons for changes in human behaviors can be very, very difficult to know with any certainty.

As a species, we seem to love causal explanations. We can see that in the way we tell stories, in our analyses of these stories, and in the grand theories we build up from these analyses. We see it in our interactions with our doctors, our therapists, our priests, our aches and pains, our TV remote controls, our meteorologists, and our children (*especially* when we're trying to get our babies to sleep). We also see it in this marvelously varied volume of studies.

Although we historians as a tribe tend to be extra cautious about offering causal explanations, like most humanists—like most *people*—we love a good causal explanation as much as any scientist. And yet, all humanists, including historians, have tended to be highly suspicious of biologists coming to offer "assistance" with causal explanations of human feelings and actions.

A few decades ago, that fear came dressed up as dire warnings about slippery slopes towards genocide and eugenics—admonitions that biological accounts of human behaviors would lead us as a species toward fascism, injustice, and the end of humanity as we have known it. These days, the suspicion is less alarmist, and also perhaps a little more self-focused: we humanists are less worried sociobiological explanations will lead to disaster, but we also don't like the idea of having the objects of our attentions "reduced" ultimately to chemicals, as if our subjects are animal bones to be cooked down for somebody else's tasty soup.

To tell many humanists that the subjects of our attractions are "reducible" to biology (and then to chemistry, and then to physics) is, I think, heard as telling us that we don't know our own work. So, many humanists are understandably hostile. The trick is, then—and I think the trick is pulled off in several places in this volume—to get humanities scholars to understand that the "consilient" perspective inspired by E.O. Wilson can help us pull out interesting *questions* rather than pushing on us boring *answers*. The consilient perspective, instead of simply splitting us into atoms, actually encourages some lumping; it asks us to try to understand what one human has in common with another as evolved beings.

Of course, willingness to entertain a consilient perspective requires a certain orientation towards empiricism, something not all contemporary humanists enjoy. (And I mean that in all the senses of "enjoy.") Some humanists don't want to hear that we should be checking our claims against the real world, and limiting our causal claims to things that can actually be checked. They lean more toward the artistic (idiosyncratic) side of the spectrum than the scientific (generalizable). What we do about that perhaps-irreconcilable worldview remains an open question, one that looks like it will unfortunately be settled by CPAs instead of PhDs.

The negative reaction some humanists have to consilience also comes, I guess, from a reading that sees the consilient approach as arrogant—as aiming at an impossible omniscience, if not omnipotence. Certainly, when the approach leads to a sprouting of just-so stories, a certain level of intellectual impatience is justified. But this isn't just your father's just-so stories. A close look at some of the essays in this volume reveals how a consilient orientation can actually foment a rather compelling level of intellectual *humility*—a recognition that the weakness of some claims in the humanities in turn expose weaknesses of some in the sciences. We all struggle with finding the causal, and we all seem to let colleagues in our own disciplines get away with causal claims much too easily.

Reading this work, I found myself realizing that, when trying to get humanists to sit at the same table in the cafeteria with the biologists, "reduction" is probably not the best way to talk about the great chain of causality. Humanists are not being naive when they find the idea of "reduction" a poor way to represent the reality of our materiality as mortals. Yes, it's surely true that physics explains chemistry explains biology; but it's also as true that, although DNA makes proteins and proteins make cells and cells make organs and organs make us, if you take away the us, our DNA dies out. The links on the chain all make the chain, so to talk about "reduction" is to pay more attention to size and to time than to what really matters to most of us in our own spatial plane (kinship, getting laid, puppies).

Moreover, where one locates the point of intellectual *satisfaction* in the study of the great chain of causality depends on what one loves to know. I love nature and I love science—that's why I'm a historian of *science*—but I will confess I love particular human histories more. I have no doubt I am an evolved animal ultimately made up of atoms. But I also have no doubt that studying physics isn't going to tell me what I really want to know about, say, the experiences and ideas of Ben Franklin, Margaret Mead, or my great-grandmother. I am not a humanist because I suck at math.

Consilience aims for a grand nonfictional story, one that subsumes all other stories and even resolves them. But the smaller stories, and our gently and vigorously causal stories about the smaller stories, seem to have a purpose. Agricultural theory is grand and important, but you still have to farm to eat. The use of stories seems so pervasive in humanity, stories must (as several authors of this volume suggest) have some importance to the species. It seems very likely the smaller stories—including the ones about Franklin and Mead and our nobody ancestors, the ones constructed in poetry, song, paint, history journals, and even the deadly prose of postmodern literary criticism—sustain or heal. Perhaps they sustain or heal individuals' psyches; almost certainly, they sustain and heal human relationships.

We could say, then, that consilience, which promises to explain all our fictional and nonfictional stories in one grand nonfictional metastory, is the most important. It's at the top, right? But we haven't needed *it* to survive this long, the way it seems perhaps we have needed all the little tales. Again, agricultural theory is grand and important, but you still have to farm to eat. Paradoxically, if Wilson's vision of consilience does anything by virtue of having science finally take the humanities seriously as its subject, it seems to verify the importance of the humanities to the species. (And I think Wilson would be delighted with that.)

Which then, in turn, would seem to mean that consilience tells humanists they have a certain importance about which they've perhaps been slacking. An obligation, a duty, a role that ought to be—maybe?—a little more focused on the rest of the people, and less on us?

In the end, this volume leaves one with both an unsettling and a liberating thought: that although we may come to understand ourselves very well as a species, we may never really fully understand ourselves as individuals. In the humanity that is our mutual cause, as *n*'s of 1, we are only ever correlations to each other. We need the grander view that is both more microscopic and more macroscopic to know why we do what we do—to know why the child in the backseat wants to know why, and why he needs also to make his tired parents laugh.

ACKNOWLEDGMENTS

Most of the contributors to this volume participated in a conference held at the University of Missouri, St. Louis, in spring 2012: Consilience: Evolution in Biology, the Human Sciences, and the Humanities. Ronald Yasbin, Professor of Biology and Dean of the College of Arts and Sciences, proposed and funded the conference.

CONTRIBUTOR LIST

Christopher Boehm
Department of Biological Sciences and Anthropology
Goodall Research Center
University of Southern California
Los Angeles, CA

Brian Boyd
English, Drama, and Writing Studies
University of Auckland
Auckland, New Zealand

Joseph Carroll
Department of English
University of Missouri, St. Louis
St. Louis, MO

Mathias Clasen
Department of English
Aarhus University
Aarhus, Denmark

Ellen Dissanayake
School of Music
University of Washington
Seattle, WA

Alice Dreger
East Lansing, MI

Herbert Gintis
Santa Fe Institute
Santa Fe, NM

Jonathan Gottschall
Department of English
Washington & Jefferson College
Washington, PA

Henry Harpending[†]
Department of Anthropology
University of Utah
Salt Lake City, UT

John Hawks
Department of Anthropology
University of Wisconsin, Madison
Madison, WI

John A. Johnson
Department of Psychology
The Pennsylvania State University
State College, PA

Daniel Kruger
Department of Health Behavior & Health Education
University of Michigan School of Public Health
Ann Arbor, MI

Dan P. McAdams
Department of Psychology
Northwestern University
Evanston, IL

[†] Henry Harpending unfortunately passed away before this book went to press.

Barbara Oakley
School of Engineering
Oakland University
Rochester, MI

Massimo Pigliucci
Department of Philosophy
The City College of New York
New York, NY

Michael R. Rose
Department of Ecology & Evolutionary Biology
University of California, Irvine
Irvine, CA

Catherine Salmon
Department of Psychology
University of Redlands
Redlands, CA

David Sloan Wilson
Department of Biological Sciences and Anthropology
Binghamton University
Binghamton, NY

Edward O. Wilson
Department of Organismic and Evolutionary Biology
Harvard University
Cambridge, MA

INTRODUCTION

Joseph Carroll

The Content and Purpose of This Volume

The term *consilience* in its modern usage was established by Edward O. Wilson's 1998 book *Consilience: The Unity of Knowledge.* Wilson's thesis had two parts: that nature forms a unitary order of causal forces, organized hierarchically, and that scientific knowledge, because it delineates nature, also forms a unitary order. This volume is designed to give an account of consilience in one major range of knowledge—the range that extends from evolutionary biology through the social sciences to the humanities. Bringing together cutting-edge scientists and scholars in all three areas makes it possible to see how far we have come toward unifying knowledge about the human species, what major issues are still in contention, and thus what areas of research are most likely, in the near future, to produce further progress.

The essays in this volume raise and give substantial answers to questions such as these: What is the precise arc of human evolution? What were the main factors driving the evolution of the human brain and human motivational system? How closely does life among contemporary hunter-gatherers mirror conditions of ancestral life? In what ways have genes and culture co-evolved, reciprocally influencing one another? How does selection at the level of individuals interact with selection among groups? How complete and adequate are our current models of human nature? How well do these models integrate ideas about human universals, individual identity, and specific cultures? How well can we now delineate the causal chains leading from elementary principles of evolutionary biology to specifically human forms of social organization, individual identity, and imaginative culture? Are human proclivities to make and consume works of art by-products of adaptations, or are they themselves adaptations? Can evolutionary thinking guide us in giving close analytic and explanatory attention to individual works of art?

In this introduction, we shall first look closely at what consilience means, then consider the particular topics of the various essays: human evolution, human nature, social dynamics, art, and narrative.

Consilience as a Theme

What Consilience Means

To say that nature forms a unitary order of causal forces, organized hierarchically, is to say that all complex phenomena can be reduced to relations among simpler elements. Ecosystems can be reduced to interactions among organisms within a physical environment. Organisms can be reduced to organ systems or, for single-celled organisms, molecular interactions inside the cell. Organs are reducible to particular kinds of cells related functionally to one another. Cells consist of components such as membranes, nuclei, and organelles. All the parts of a cell are compounded of specific molecules, and specific molecules are formed by bonding among chemical elements. The chemical elements are atoms with specific numbers of protons, neutrons, and electrons. Protons and neutrons are, in turn, composed of subatomic particles. At levels of analysis available now only to informed scientific speculation, particles yield to still more basic structures such as strings.

It is in the nature of analysis to reduce complex structures to simpler elements. This process can, of course, be reversed. We can begin with subatomic particles and observe the way more complex structures emerge at higher levels of organization among component parts. Atoms interact to form molecules; organic molecules combine to form DNA, which regulates the organization of other molecules into cells and physiological processes. Cells combine to form tissues and organs; organ systems form organisms, which sometimes form social groups. Organisms interact with each other and with the physical environment to form ecosystems. The elementary components in all higher levels of organization are themselves composed of still smaller components at lower levels in the causal hierarchy. For instance, the elementary components in a social group are individual organisms, but individual organisms are themselves complex systems of organs or, for single-celled organisms, organelles and molecules.

The causal hierarchy in the natural order has emerged over time. Two main fields delineating this emergence are cosmological physics and biological commentary on "major transitions" in the evolution of life. Cosmological physics begins with the Big Bang and describes the formation of atoms, stars, and solar systems (Weinberg 1992, 1993). Commentaries on the "major transitions" of life begin with the formation of self-replicating molecules and work up through ever-more complex levels of organization—from nonnucleated to nucleated single-celled organisms, multicellular organisms, organisms with organ systems, social animals, and human cultures (Maynard Smith and Szathmáry 1995; Shubin 2008; Lane 2009; Bourke 2011; Pross 2012; Shubin 2013).

So long as we are discussing only the physical and biological range of knowledge, most educated people would agree that scientific disciplines form a causal hierarchy corresponding to levels of organization in the physical world. Physics and astronomy deal with the fundamental forces in the physical world—gravity, electromagnetism, and strong and weak nuclear forces. Chemistry as a discipline begins with the organization of subatomic particles into specific kinds of atoms—the chemical elements. Geology explains how these elements have been organized in the history of Earth. Biology begins with the organization of chemical elements into organic molecules and explains the development of life through natural selection. Thus far, "consilience" seems little more than educated common sense.

It is only when we get to the human world that disagreement begins. Many educated people still maintain various forms of human exceptionalism—the idea that the human mind or spirit, or human culture, somehow stands apart from the causal hierarchy that prevails in the rest of the natural order. Almost everyone would agree that the human world introduces something new to reality. From the consilient perspective, it is a "major transition," a more complex organization of the elements that produce atoms and chemicals, organisms and ecosystems. The alternative is that the human world is not merely a major transition but, rather, a qualitatively different kind of thing that separates humans fundamentally from the physical world. A nonconsilient perspective is thus necessarily dualistic. From a dualistic perspective, one part of the world consists of physical elements that combine into more complex forms of organization, which are, conversely, reducible to their components. That physical world is accessible to science. From the dualistic perspective, the other part, the spiritual or cultural part, can perhaps be influenced by physical elements, but it can never be reduced to those elements.

The consilient worldview is monistic. Researchers adopting this worldview do not believe that human mental experience gives evidence for any peculiarly human stuff that cannot be reduced to interactions among components in the physical world. From the monistic perspective, imaginative culture—norms, religious beliefs, ideologies, philosophies, and the arts—are products of brains interacting with other brains and are thus reducible to electrochemical interactions among neurons. Brains are embedded in environments, both social and physical. Cultural traditions form major elements in social environments, but cultural traditions are themselves the products of brain activity among social organisms transmitting information by means of symbols.

While advancing steadily since the Renaissance, science has approached ever closer to the human realm, advancing from astronomy ("celestial mechanics") to chemistry, anatomy and physiology, to geology and biology. During the past 40 years, the social sciences have finally taken a decisive turn toward consilience. Evolutionary social science began with the publication of Huxley's *Evidence as to Man's Place in Nature* in 1863, followed up by Darwin's *Descent of Man* in 1871, but in the second decade of the 20th century, anthropology and sociology segregated themselves sharply from evolutionary biology. The idea that culture is an autonomous human order, detached from evolved and genetically transmitted dispositions, governed standard social science from about 1911 through the 1970s (Degler 1991; Pinker 2002). The new Darwinian revolution of the past several decades is not yet complete, but has clearly passed the point of no return. One important indicator for this tectonic shift can be discerned in commentary on human behavior aimed at educated general readers. For the past several years, much of the most successful such commentary—successful in terms of sales, reviews, and prestige—has been deeply versed in current knowledge about evolved dispositions and neurobiological mechanisms such as brain structures, hormones, and neurotransmitters (Ridley 1994, 1996; Pinker 1997, 2002; Buss 2005b; Goleman 2006; Haidt 2006; Wade 2006; Angier 2007; Linden 2007; Cacioppo and Patrick 2008; Carroll 2009; Cochran and Harpending 2009; Dutton 2009; Lane 2009; Wrangham 2009; Damasio 2010; Thagard 2010; Baron-Cohen 2011; Brooks 2011; Kean 2011; Kenrick 2011; Linden 2011; Pinker 2011; Gottschall 2012; Haidt 2012; Kean 2012; Wilson 2012). Currently, writers who ignore or deprecate biological influences on behavior would, in the judgment of many generally educated readers, relegate themselves automatically to the margins of informed discussion.

Among academic disciplines, the humanities are the strongest outpost of resistance to a monistic worldview grounded in evolutionary biology. Humanists are heavily dependent on theories in other disciplines but make little or no use of empirical research in other disciplines. In the academic literary establishment, as in other humanistic disciplines, the current framework of theory is still grounded in obsolete forms of sociology (Marxism), psychology (Freudianism), and linguistics (Saussurean linguistics and Derridean language philosophy) (Carroll 1995). For the first three-quarters of the 20th century, humanists treated the arts as the prime medium for the supposedly transcendent autonomy of the human spirit. For the past four decades, "discourse" or "culture" has been invested with autonomy (Abrams 1997; Carroll 2011b, 259–277). For about two decades now, though, evolutionists in the humanities have been making sustained progress in incorporating information from empirical research in the evolutionary social sciences and have sometimes also incorporated empirical methods. Although evolutionary humanists like those included in this volume still constitute only a tiny fraction of academic humanists, some anticipate that the humanities as a whole will eventually be fully assimilated to the consilient worldview.

Because nature is continuous in organization, disciplines focused on specific levels in the causal hierarchy bleed into each other at the margins. Chemists dealing with nanostructures often do work indistinguishable from that of physicists. Biochemists and molecular biologists occupy the border ground between chemistry and biology, and, indeed, most biologists necessarily deal with phenomena at a chemical level. Neurobiologists, for instance, occupy themselves with the chemical components of hormones and neurotransmitters. Biologists concerned with animal behavior are "ethologists," and one branch of ethology is "human ethology" (Eibl-Eibesfeldt 1989). Primatologists compare species ranging from lemurs and baboons to chimpanzees to humans. Some of the most important findings about human social evolution and human cognitive development have been produced by researchers who compare chimpanzees and humans, thus working simultaneously as primatologists and anthropologists or primatologists and cognitive scientists (Boehm 1999; Tomasello et al. 2005; Boehm 2012). In developing hypotheses about human social and cognitive evolution, anthropologists and archeologists extend their reach backward to hominins and forward to the Neolithic (Mellars and Stringer 1989; Mithen 1996; Klein 2002; Mithen 2004; Wade 2006; Mellars 2007). Genetics, anthropology, and cultural history converge in the study of gene-culture coevolution (Richerson and Boyd 2005; Cochran and Harpending 2009). Scientists with a primary training in fields such as biology, archeology, or psychology have developed hypotheses about the human proclivity for producing aesthetic ornamentation and works of imagination (Darwin 1871; Mithen 1996; Wilson 1998; Miller 2000). Humanists with a primary training in fields such as philosophy, art history, or literature have assimilated and critiqued those hypotheses (Dissanayake 2000; Boyd 2009; Dutton 2009; Boyd et al. 2010; Carroll 2012; Gottschall 2012).

Biology is the pivotal discipline linking the physical sciences, the social sciences, and the humanities. The building blocks of biology are derived from chemistry, geology, and even directly from physics—for example, the influence of planetary motions on the diurnal rhythms of plants and animals, the effects of sunlight and other sources of thermal energy on the life cycle of individual organisms and ecosystems, and the way animals use light, sound, and electromagnetism for navigating their environments. In the other direction in the causal hierarchy, basic concepts in evolutionary biology inform virtually all evolutionary research in the social sciences and the humanities: adaptation by means of natural selection and sexual selection (Darwin 1859, 1871), inclusive fitness (Hamilton 1964a, 1964b), differential parental investment (Trivers 1972), and reciprocal altruism (Trivers 1971).

Because nature forms a causal hierarchy, influence is asymmetrical among disciplines at different levels in the hierarchy. Discoveries in physics are more likely to influence chemistry than discoveries in chemistry to influence physics, and discoveries in chemistry are more likely to influence biology than the other way around. So also for biology and the social sciences, and for the social sciences and the humanities. Researchers in disciplines downstream in the causal hierarchy nonetheless have a crucial role in determining whether efforts at causal reduction wrongly strip out emergent phenomena in their own fields. For example, social scientists have effectively countered the idea that all human social interaction can be reduced simply to reciprocal altruism and have formulated more adequate alternative hypotheses (Haidt 2012). Humanists have effectively countered the idea that all human imaginative production can be reduced to sexual display and have formulated more adequate alternative hypotheses (Dissanayake 2000, and Chapter 7 in this volume; Carroll 2008a, 119–128, 2008b, 349–368, 2012).

Researchers in disciplines downstream in a causal hierarchy are not solely dependent on causal explanations from upstream disciplines. Good explanations identify simpler forces at work in complex phenomena, but emergent structures also have causal force on one another. For instance, populations within ecosystems have reciprocally causal effects on one another and on individual organisms within each population (Darwin 1859; Wilson 1992). Large-scale institutional structures—nation states, political parties, economies, and religions—interact in reciprocally causal ways with each other and with the evolved psychological characteristics of individual people (McAdams 2006; Turchin 2006; Haidt 2012). So also with subjects in the humanities. Artistic traditions and conventions have reciprocally causal effects on each other and on individual artists and individual works of art (Wilson 1931; Wellek 1949a, 1949b; Abrams 1953; Wellek 1961; Abrams 1965; and see Boyd, Chapter 13, in this volume). Good explanations at any level of emergent complexity are likely to identify causal relations among forces at that level and link them with causal forces at lower levels in the causal hierarchy.

Good explanations take account of the causal interactions among emergent phenomena, but valid conceptions of emergent phenomena depend on correctly identifying the elements that make up the emergent phenomena (Pinker 2005). Ignoring principles of natural selection, for instance, produces false conceptions of the way populations interact with each other and with individual organisms (Carroll 2001; Easterlin 2004). Deprecating or dismissing ideas of an evolved and adapted human nature produces false conceptions of the way institutions and cultural practices interact with each other and with individual people (Freeman 1983; Degler 1991; Tooby and Cosmides 1992; Pinker 2002). Repudiating the idea that authors intend to communicate definite meanings about a shared reality produces false conceptions of the way literary conventions interact with each other and with individual artists and their audiences (Carroll 1995; Abrams 1997; Boyd 2006).

Specialists can give expert testimony in their own fields, but there are no rigid boundaries in a consilient research community. Biologists also have intervened effectively in expanding the scope of concepts included in the analysis of human social dynamics (Sober and Wilson 1998; Wilson 2012). Biologists and social scientists have contributed in important ways to evolutionary theories about religion, the arts, and other products of human mind and imagination (Mithen 1996; Wilson 1998; Wilson 2002; Dissanayake 2011).

Researchers downstream in a causal hierarchy sometimes express resentment at the idea that they are on the receiving end of an asymmetrical disciplinary influence. Such resentment is hardly a scientific motive, but it is a human fact, and it has had and will perhaps continue to have a distorting influence on intellectual history. When the social sciences, at their inception, segregated themselves from biology and declared their independence, they were moved in part by a determination to focus on organizational principles appropriate to their particular fields of study. But they were also moved by a desire to assert causal primacy, at whatever cost to logic or explanatory power (Fox 1989; Degler 1991; Pinker 2002). Similar motives, both good and bad, can often be detected among humanists resisting connections to biology or the evolutionary social sciences (Dawson 2006; Goodheart 2007; Deresiewicz 2009; Kramnick 2011).

As a counterweight to that kind of biasing resentment, it is wholesome to remember that a hierarchy of causal reduction can be flipped over into a hierarchy of emergent complexity. Chemists absorb the principles of physics and introduce new causal principles active in the relations among chemical elements; biologists absorb chemistry and deal with phenomena—organisms and ecosystems extending over evolutionary time—at levels of complexity higher than that of the periodic table. Evolutionary social scientists absorb the fundamental principles of biology but also study forms of human social organization that are more complex than the social organization of other species. Evolutionary humanists, working downstream from all these disciplines, absorb their explanatory principles but also deal with the products of the human mind, with religion, myth, philosophical speculation, and cultural history, and with works of art. Following the logic of emergent complexity, the topics that are the peculiar province of humanists are the most complex subjects available to scientific inquiry—so complex that for a majority of humanists still, and for not a few scientists, the gap between the sciences and the humanities presents itself as an unbridgeable gulf.

Causal reduction and emergent complexity are the two poles of a consilient universe. For researchers alert to the continuum between those two poles, the boasting rights of either pole are far less important and interesting than the advances in knowledge that can be produced, in any given field, by delineating linkages between them. To give a salient example, the theory of gene-culture co-evolution is now in its infancy but is clearly a central point of convergence for biology, the social sciences, and the humanities. Within the next 20 years, it seems likely that research in this area will produce some of the most important advances in an evolutionary understanding of the human species. Those advances will depend on work that synthesizes findings in all three major areas of research. Geneticists and evolutionary biologists are in the best position to identify changes in gene frequencies relevant to human social and cultural activity. Social scientists, including anthropologists and archeologists, are in the best position to delineate the elementary forces at work in human social organization over both evolutionary and historical timescales. Humanists are in the best position to identify the character and structure of the products of the imagination—religions, ideologies, stories, music, and the visual arts—that interact in reciprocally causal ways with the evolved dispositions commonly designated by the term "human nature." Researchers in any of these three areas can assimilate findings from the other two areas, characterize the current state of knowledge, and generate new hypotheses that stimulate further research.

Challenges to the Idea of Consilience

Two of the essays in this volume, those of Hawks and Pigliucci, express skepticism about the possibilities of consilience. Hawks comments on the difficulty of being able to identify empathically with the subjective lives of ancient peoples, with special reference to Neanderthals. Empathic identification would require the anthropologist to cross boundaries between species-typical forms of sensation and also between radically different ecological conditions. Working out from this problem, Hawks draws a broad contrast between scientific and interpretive schools or styles of anthropology. The scientific style uses meticulous reconstruction of objective facts and seeks causal reductions empirically tested by predictions. The interpretive style, as Hawks conceives it, plays with broad speculative theories, bringing them into conjunction with the subject matter of anthropology. Interpretive anthropologists "focus on the aesthetics of an improvised encounter between observation and theory." Despite such conflicts in standards and values, Hawks believes that "we can develop some knowledge about the subjective lives of these people." Empirical inquiry and speculative theory are always in some tension, but the tension can be productive. Hawks affirms that in using scientific methods we are also "building a humanistic understanding of Neandertals and other ancient people."

Hawks's essay probes potential methodological problems in achieving consilience. Pigliucci's essay, in contrast, suggests that consilience, as a philosophical vision and a research program, is essentially misconceived, so that its fulfillment is neither possible nor desirable. As a scholar and scientist with doctoral degrees in genetics, botany, and philosophy, Pigliucci is in a good position to offer a representative sampling of objections to the idea of consilience. The version of consilience that Pigliucci criticizes is reductionist in purpose but has little concern for "convergence" among disciplines, it fails to register the emergence of more complex structures from the interaction of simpler elements, it is committed to a strong version of "meme" theory—the idea that bits of cultural information are essentially parallel in character and function to genes—but overlooks the theory of gene-culture coevolution, and it pursues "ultimate objective truth," an unattainable will o' the wisp, as its chief goal. One might reasonably question whether that version of consilience is represented by any actual person. In any case, it is represented by none of the essays in this current volume.

As an alternative to the consilient program, Pigliucci suggests that the traditional division of the disciplines, especially the divide between the humanities and sciences, represents a natural and necessary accommodation to the human mind as it has developed historically. "There may be better ways to organize our knowledge, in some absolute sense, but likely what we have come up with is something that works well for us as biological–cultural beings of a certain type and with a certain history." Other contributors to the volume regard the current arrangement of academic disciplines as a historical artifact that reflects obsolete conceptions of life and mind. Wilson, for instance, describes the current arrangement not as a culminating and final state of affairs but as a transition to a more complete and unified vision of human life:

> The major features of the biological origins of our species are coming into focus, and with this clarification the potential of a more fruitful contact between science and the humanities. The convergence between these two great branches of learning will matter hugely when enough people have thought it through. On the science side, genetics, the brain sciences, evolutionary biology, and paleontology will be seen in a different light. Students will be taught prehistory as well as conventional history—the whole presented as the living world's greatest epic.

Wilson's formulations imply we already have the information necessary for this transformation. The only remaining obstacle is that enough people have not yet "thought it through." If Wilson is correct, it seems likely that this one remaining obstacle will be rapidly eroded by the intellectual vigor of the species, the increasing speed at which knowledge is generated, and the expanding means for the sharing of knowledge.

Degrees of Paradigmatic Consensus in the Disciplines

A scientific paradigm is a stable framework within which researchers can produce progressive, cumulative knowledge. The framework is stable because researchers agree that its core concepts are internally coherent, grounded in empirical findings, and concordant with concepts in other empirically grounded disciplines. It is progressive because its core concepts are so broad and basic they can incorporate new discoveries. In geology, for instance, the paradigmatic synthesis produced by Charles Lyell during the 1830s was so broad and basic that it could incorporate 20th-century discoveries about plate tectonics.

During the 1930s, the period of "The Modern Synthesis," evolutionary biology finally achieved the status of a paradigm. Darwin had provided basic materials for that paradigm in the theory of descent with modification by means of natural selection, but uncertainties about the mechanisms of inheritance rendered the theory of natural selection controversial for some seven decades after the theory had first been proposed (Huxley 1942; Mayr 1982; Bowler 1988). The Modern Synthesis had sufficient breadth and validity so that it could be expanded, during the next several decades, to include discoveries such as DNA, inclusive fitness, differential parental investment, and, most recently, multilevel selection, including selection at the level of the group.

Evolutionary psychology is still in the process of forming a paradigm. Early sociobiology too directly invoked the theory of fitness maximization as a primary motive in human behavior. Evolutionary psychologists corrected that mistake by insisting on an intermediate stage of proximal mechanisms—for instance, the desire for sex rather than the desire for offspring (Symons 1992). In turn, though, the early evolutionary psychologists eliminated or minimized the domain-general powers of human intelligence (Mithen 1996; Sterelny 2003; Geary 2005), failed to register the systemic relations among disparate proximal mechanisms (Smith et al. 2001; Kaplan and Gangestad 2005), eliminated or minimized the significance of individual variation in humans (Nettle 2006, 2007), oversimplified the environment of evolutionary adaptedness EEA (Foley 1995; Irons 1998; Potts 1998), envisioned an exaggerated contrast between modern conditions and human adaptive dispositions (Zuk 2013), and thus forestalled any adequate recognition of ongoing gene-culture coevolution (Richerson and Boyd 2005; Cochran and Harpending 2009). During the past two decades, evolutionary social scientists have been correcting all those premature theoretical reductions and thus building more accurate models of human evolution and human nature. During this same period, the early sociobiological emphasis on selection at the level of individuals has been giving way to a more complex understanding of evolved human dispositions for social life (Sober and Wilson 1998; Boehm 1999, 2012; Haidt 2012; Wilson 2012). Evolutionary psychology has not yet fully stabilized as a true paradigm, but it is well on the way (Buss 2005a; Dunbar and Barrett 2007; Gangestad and Simpson 2007; Laland and Brown 2011). The most important phenomena that have yet to be fully incorporated within a reasoned consensus are the products of imaginative culture—the arts, religions, philosophies, and ideologies (Dissanayake 2000; Boyd 2009; Boyd et al. 2010; Carroll 2011b; Dissanayake 2011; Gottschall 2012).

Several of the essays in this volume discuss active controversies within their own fields. Wilson explains the conflict over group selection. Boehm surveys the various hypotheses that have been proposed to explain altruism and makes a case for a comprehensive theory that incorporates and revises existing hypotheses. Harpending and Harris propose serious qualifications for the common assumption that hunter-gatherers can be taken as proxies for ancestral humans. All these issues are important, but they are all also points of dispute within a broad consensus about the evolved and adapted character of the human mind. Contributors from literary studies register more basic disagreements within their own discipline. Focusing specifically on horror fiction, Clasen sets his evolutionary approach into sharp contrast with "theoretically flawed approaches that have dominated horror studies in recent decades, especially psychoanalysis and the various forms of 'blank slate' political ideology." In a similar vein, Boyd contrasts a "biocultural" approach to literary study with the exclusively cultural approach that has dominated literary study for decades. Carroll et al. observe that the majority of literary scholars still reject an evolutionary view of human behavior and, even more broadly, the idea that science can produce objective knowledge. Carroll et al. argue that literary studies currently display the kind of "epistemic disorder that characterizes most disciplines in the period before a paradigm has formed."

Cross-Disciplinary Linkages

Oakley observes that "the history of science has shown that it's possible to work for decades—even centuries—using an underlying perspective on a given challenge or situation that makes it impossible to make progress." Such situations require reframing the problem, changing the context of inquiry. As Oakley puts it, "a subtle, simple perspective shift can allow for vital breakthroughs to take place." Oakley is one of several contributors who link multiple disciplines either to produce broad general ideas or give concentrated analytic attention to particular topics. Linkages include biology and paleoanthropology (Wilson), psychology and engineering (Oakley); paleoanthropology and developmental psychology (Dissanayake), and psychology and narrative theory (McAdams; Carroll et al.; Boyd; Clasen).

Wilson's special areas of expertise include entomology and the evolution of sociality across the animal kingdom ("sociobiology"). Integrating information from that range of expertise with information from paleoanthropology and hunter-gatherer culture, Wilson develops a general theory of "eusociality." That theory, zoological in scope, has profound implications for the way we envision specifically human forms of social behavior. It creates a new context of inquiry for the ecological conditions underlying group formation, the division of labor, the relations among generations, and the relations between discrete social groups.

Oakley describes the intellectual path that led to her book *Evil Genes* (2007). By integrating research in neuroscience, personality disorders, and cultural and political history, she has been delineating the continuum between individual psychopathology and pathological social organization at the level of institutions and nation states. In her essay for this volume, Oakley brings her interdisciplinary expertise to bear on the problem of "pathological altruism." Discussing the cognitive bias that leads to dysfunctional forms of altruism, she makes a case that engineering could help put social and psychological theories to the hard test of reality. She sketches out a practicable interdisciplinary program in psychology and engineering. Neuroimaging points toward the kind of mechanization that has made such a profound difference in medical science. Connecting engineering with psychology extends the range of consilience, in this volume, beyond the life sciences. The mechanics of neuroimaging are grounded in physics as well as in neurochemistry.

Dissanayake, Boyd, and Clasen bring multiple fields to bear on highly specific topics in imaginative culture. Dissanayake's topic is mark-making among preliterate peoples: cupules and engraved or painted geometrics. To make sense of this topic, she incorporates ideas from anthropology, archeology, developmental psychology, neuroscience, and ethological research on a contemporary aboriginal group. Her chief disciplinary affiliations are human ethology and developmental psychology. She synthesizes current thinking on ancient petroglyphs and pictographs and uses developmental cognitive psychology to delineate parallels between ancient mark-making and the mark-making of young children. All this highly particular information has implications for a much broader issue: the controverted question of "the human revolution"—that is, the timing and pace at which humans began to produce distinctively human imaginative culture. Clasen's topic is a specific genre of fiction in literature and film: horror. Why do people love to scare themselves with fictional monsters? To answer that question, Clasen synthesizes information from paleoanthropology, social psychology, and affective neuroscience. His conclusions make it possible for him to contribute evidence toward a crucial issue in human evolutionary theory—whether the arts are adaptively functional components in gene-culture coevolution. Boyd's topic is a set of parallels in processes in evolution, the sciences, and the arts. He uses two basic heuristics—problem–solution and cost–benefit—to provide a flexible analytic model for commentary on poetic and narrative structures. Drawing on cognitive and social neuroscience, he frames specific works of art as instances of "pattern recognition" and "shared attention." Like Clasen, he brings his conclusions to bear on the question of the adaptive function of the arts.

McAdams and Carroll et al. use overlapping bodies of interdisciplinary information to formulate complementary ideas about human identity and "meaning" in narrative. They integrate information from multiple fields of psychology to construct comprehensive models of individual identity. McAdams uses narrative theory to illuminate the autobiographical narratives of real individual people. Carroll et al. use the psychology of real individual people to illuminate the construction of fictional characters in Victorian novels.

McAdams's home discipline is personality psychology, but he has expanded the scope of personality psychology to include more than the usual five-factor set of personality variables (extraversion, conscientiousness, agreeableness, neuroticism, openness to experience). From evolutionary social psychology, he derives a set of basic life goals. Integrating developmental, cognitive, and narrative psychology, he constructs a theory of autobiographical narrative, the story every individual tells about his or her life. A life narrative is the way humans make "meaning" out of their lives. Life narratives delineate a continuously unfolding identity in which main sequences are shaped not just by events and actions, but also by goals achieved or not achieved, values affirmed or subverted, needs fulfilled or frustrated. Life narratives include the largest contexts within which people locate their own individual self-images—families, friends, communities, the natural world, and, for many people, a spiritual world. Meaning in a life derives largely from the value with which those contexts are invested—love, devotion, awe, reverence, pride—and the value attributed to one's place in them, whether one is loved or detested, despised or admired. Sustaining or changing the image of oneself within such contexts forms a chief motive for behavior.

Carroll et al. use human life-history theory to construct a set of basic motives and use Ekman's theory of basic emotions to register readers' emotional responses to characters. They differentiate individuals through motives and personality traits (the five-factor system) and differentiate sexes through motives and criteria for selecting mates. By correlating features of identity in characters with the valenced emotional responses of readers to the characters, Carroll et al. identify the structure of values that prevails across the whole body of novels. By having readers sort characters into protagonists, antagonists, and minor characters, they produce a synoptic image of the positive and negative values that are shared by the authors and their readers. They report that antagonists are chiefly motivated by a desire for dominance. Protagonists are heavily motivated by self-effacing prosociality. That valenced antithesis helps form a community of shared values within which authors and readers construct their own life narratives.

Carroll et al. use empirical, quantitative methods. Two of the team members (Carroll and Gottschall) have training primarily as literary scholars, and two (Johnson and Kruger) primarily as psychologists. In interpreting their results, they invoke Boehm's claim that hunter-gatherers suppress dominance in individuals. They thus construct a model of human nature from concepts in biologically grounded psychology, use that model to produce empirical data about a literary subject, and interpret that data with ideas from research in primatology and anthropology.

The Trajectory of Human Evolution

Several contributors converge on two main themes in current thinking about human evolution: multilevel selection and gene-culture coevolution. Wilson, Boehm, McAdams, and Carroll et al. invoke selection pressures at the level of social groups. Wilson, Boehm, Harpending and Harris, Rose, McAdams, Clasen, Dissanayake, and Carroll et al. delineate ways in which ecological or social conditions have interacted causally, over evolutionary timescales, with specifically human forms of intelligence and imagination. Within that broad convergence, the contributors display significant differences of focus and emphasis.

Wilson and Boehm both identify hunting and meat-sharing as main factors in human evolution. Wilson also designates the use of defensible campsites as a pivotal event in human evolutionary history. Organizing social life around a campsite, he argues, generates a self-perpetuating cascade in human social intelligence. He postulates a causal connection between human social intelligence and the evolution of a specifically human power "to invent and rehearse competing scenarios." Wilson and McAdams both identify "group selection" as a major evolutionary force among humans. Boehm includes "group selection" in a list of the main theories that have been proposed to account for "altruistic" behavior in humans, but Boehm identifies a form of "social selection" that is distinct from "group selection" and interactive with it. Presenting evidence from modern hunter-gatherer populations, Boehm argues that specifically human forms of cooperative behavior depend crucially on suppressing dominance behavior in individuals. Social pressure constraining individuals to suppress dominance behavior selects for the ability to internalize group norms, an ability that leads, ultimately, in modern humans to a specifically human form of social imagination: ideology. McAdams and Carroll et al. reflect on the way narratives display internalized group norms.

Future research on human evolution will almost certainly concentrate on the tension between conserved adaptations and novel genetic attributes. In this volume, Wilson, Boehm, Dissanayake, Clasen, and Carroll et al. give a strong emphasis to the conservation of evolved dispositions—to the persistence of adaptations from among ancestral populations. Harpending and Harris point in the opposite direction. Taking up one particular issue—the tendency of people now to give preferential treatment to ethnically similar people—they give evidence for the selective force produced by relatively recent forms of human behavior: sedentism (living in settled communities), agriculture, and the pooling of ethnically diverse people in cities. Boehm and Harpending and Harris reflect on the relatively recent and sudden emergence of culturally modern human behavior—the kind of behavior that produces complex tools and symbolic artifacts. Dissanayake, in contrast, stresses gradualism and continuity in cultural development. Rose constructs an evolutionary hypothesis that emphasizes neither conserved nor novel traits for specific forms of behavior. Instead, he identifies a set of "general-purpose brain functions" that are "useful for both ecological and social competition" and that thus help explain "the complexity and versatility of human behavior."

A Biocultural Conception of Human Nature

Human Life History and Three Specifically Human Forms of Culture

All species have a nature—an evolved set of species-typical behaviors. In all species, these behaviors form a functionally integrated suite adapted to satisfy the two basic requirements of life: sustaining a body and reproducing (Alexander 1987; Lummaa 2007). The total life trajectory or "life history" of every species is a reproductive cycle that includes particular forms of birth, development to adulthood, mating, and longevity (Kaplan and Gangestad 2005; Flatt and Heyland 2011). Sustaining life involves adaptations for acquiring food and protecting the organism from environmental threats. For mammals, birds, and some other species, evolved characteristics also include dispositions for nurturing offspring. For social species, evolved characteristics include species-typical ways of interacting with conspecifics. For eusocial species, those forms of interaction involve divisions of labor and cooperation aimed at fulfilling the needs of the group. Some few highly intelligent species transmit learned behavior that includes using simple tools.

Humans have developed the capacity for transmitting information in three ways that are either unique to human culture or exceptionally developed in it: (a) they retain and develop innovations to produce cumulative forms of learned behavior—social, mechanical, and intellectual (Sterelny 2003; Tomasello et al. 2005; Boyd and Richerson 2007); (b) they extrapolate general ideas (Hawkins 2004; Geary 2005); and (c) they produce imaginative artifacts (Dissanayake 2000; Dutton 2009; Carroll 2011b; Gottschall 2012). Through cumulative innovation, humans have transformed techniques into technology, tribes into civilizations, discoveries into progressive sciences, and individual works of art into artistic traditions. By extrapolating general ideas, they have produced ideologies, religions, philosophies, histories, scientific theories, and theories about civilization. Animals of other species dream, produce emotionally expressive vocalizations, engage in play, and even, in the case of bower birds, fashion aesthetically designed artifacts. Only humans produce imaginative artifacts through which they depict objects and actions, evoke the subjective experience of other creatures, express their own attitudes to those experiences, affirm or contest social norms, communicate systems of belief, and convey worldviews.

The three features that distinguish specifically human forms of culture—cumulative innovation, general ideas, and imaginative artifacts—interact in ways that have progressively altered the functionally integrated suite of adaptive behaviors in the hominin lineage. Gene-culture coevolution, beginning with the use of tools and the control of fire, has altered hominin characteristics all the way down to anatomy and physiology (Cochran and Harpending 2009; Wrangham 2009; Carroll 2011a). In *Homo sapiens*, culture is shaped and directed by genetically transmitted features of an evolved and adapted human nature, but cultural practices also form emergent levels of complexity in which the basic features of human nature interact with each other to produce phenotypically novel forms of behavior.

Humans are still driven by basic animal needs such as hunger and thirst, and they are still dependent, as a species, on the reproductive cycle. They have inherited from their ancestors forms of anatomy and physiology adapted to an omnivorous diet of cooked foods (Wrangham 2009); anatomical and cognitive traits derived originally from adaptations for living in trees and then for hunting and gathering on the ground (Wade 2006; Klein 2009; Boyd and Silk 2012); dispositions for pair bonding, dual parenting, and multigenerational care of the young (Geary and Flinn 2001; Wilson 2012); aptitudes for intense social interactions in groups that work cooperatively to acquire resources and defend the group from predators, including other human groups (Wilson 2007; Boehm 2012; Wilson 2012); and impulses of fear and aggression in relation to threats (Buss 2005b; Panksepp and Biven 2012; Shackelford and Weekes-Shackelford 2012). All those biological characteristics form part of every known human culture. They are "human universals" (Brown 1991). Distinct human cultures organize these universal characteristics in different ways. Humans adapt to local ecological conditions by developing traditions in technology, social organization, belief systems, and artistic practices. The common elements among these local traditions, though, are themselves human universals. All human cultures have technology, complex social organization, belief systems, and artistic practices (Brown 1991; Dissanayake 2000; Baumeister 2005). The capacity to produce such traditions are part of the genetically transmitted features peculiar to the species. Humans are thus truly a biocultural species—the only biocultural species.

Sociality and Imagination

All the essays in this volume take as their subject one or another aspect of "human nature." From these various aspects, two main themes emerge: human nature is ultrasocial and it is imaginative. Wilson offers a representative statement about human sociality. "We are compulsively driven to create and belong to groups, variously nested, overlapping, or separate, and large or small." Affirming Wilson's ideas about eusociality and group selection, McAdams declares that it is part of human nature "to identify closely with groups, for throughout human evolution, individual survival has depended on the survival of the group as a whole and, more important, on one's particular standing within the group." Wilson also offers a representative statement about the human imagination, especially in its narrative forms. "We instinctively delight in the telling of countless stories about others as players on the inner stage. The best of it is expressed in the creative arts, political theory, and other higher level activities we have come to call the humanities." Dissanayake foregrounds a more basic form of imagination: the aesthetic "primitives" that manifest themselves first in nonsymbolic forms. She argues that humans have "an evolved behavioral predisposition" to "use special devices that attract attention, sustain interest, and create and manipulate emotion. Such devices include simplification or formalization, repetition, exaggeration, elaboration, and manipulation of expectation." Dissanayake directs attention away from the elite arts that are the typical subjects of the humanities. She focuses instead on the universal character of artistic activity. "All human societies perform ceremonial practices or rituals in which several arts combine—song, dance, and dramatic storytelling, in addition to the visual panoply of costumes and other body adornment, masks, altered surroundings, and special objects, which could include painted or carved marks on stone."

Ultrasociality and imagination are intertwined in reciprocally causal ways. Several of the essays in this volume probe those causal relationships. Wilson attributes the evolution of imaginative activity directly to social interaction. Boehm, Dissanayake, McAdams, and Carroll et al. identify forms of imaginative activity that serve adaptive social functions. Clasen, invoking a hypothesis originally formulated by Wilson, explains the adaptive function of imagination in ways that encompass sociality but are not limited to it.

Research into adaptations for living in social groups has converged on one basic dichotomy, variously formulated: cooperation and competition, getting along and getting ahead, affiliation and dominance. As McAdams explains, "Going back even to Freud's (1930/1961) famous dichotomy of Eros and aggression, researchers have repeatedly distinguished between two classes of basic human motivations: those designed to promote communion, love, intimacy, affiliation, group bonding, and interdependence on the one hand; and those aimed to promote individual control, power, status, achievement, self-expansion, and independence on the other." In this volume, Wilson, Boehm, Harpending and Harris, McAdams, and Carroll et al. use variations on this dichotomy to pry open the complexities of social relationships. McAdams invokes Wilson's idea that competition between individuals within groups parallels competition between groups, but Wilson and McAdams both also acknowledge that life within groups consists of a perpetual dynamic tension between cooperation and competition. Boehm gives close attention to the way group life transforms prosocial behavior into a selective advantage for individuals within a group. Carroll et al. argue that Victorian fiction stigmatizes dominance behavior, affirms prosociality, and thus helps bind its readers into members "within a community dependent on shared norms of cooperative behavior."

Humans are so thoroughly social, and also now live in an environment so thoroughly domesticated to human use, that it is easy to lose sight of the way social life fits into the broader suite of adaptive characteristics in human nature. McAdams quotes psychologist Robert Hogan, who affirms that "getting along and getting ahead are the two great problems in life that each person must solve." From a life-history perspective, the two great problems in life are survival and reproduction. The complications of social life are built on that foundation. Wilson and Boehm are certainly correct that for the human species subsistence and sociality are closely intertwined. Cooperative hunting for meat is a core feature in the evolutionary trajectory of the species. So, too, for humans, as for all mammals, mother–infant bonding is a core feature in the trajectory of the species. For humans, mother–infant bonding is embedded in complex social networks that include sexual pair bonding, dual parenting, multigenerational cooperation in the care of offspring, and cooperative work. Without integration in a social group, ancestral human mothers would not have been able to rear offspring. Nonetheless, in the hierarchy of causal reductions leading to inclusive fitness, adaptations for social life are adaptations to aid survival and reproduction—not the other way around.

Evolutionary commentary on imaginative artifacts has centered on the question of adaptive function: whether they have any adaptive functions, and if so, what those might be. Carroll et al. summarize the various theories that have been put forward. Dissanayake gives a critique of the idea that imaginative activity serves chiefly as a form of sexual display. McAdams considers the hypothesis that the arts are essentially a nonadaptive by-product of other characteristics that have adaptive value. He argues that storytelling is grounded in universal dispositions but suggests that highly individualized autobiographical narratives are particularly salient and psychologically functional in complex modern cultures. This is an empirical question of considerable interest. In literary theory, it can be closely associated with a canonical historical issue: the rise of the novel (Watt 1957; McKeon 1987). Is it the case that a personal life narrative becomes crucially important only in modern societies? Or is it the case that all people, even those living in preliterate cultures using simple forms of technology, have a universal need to envision their own life trajectories within their total worldview? Do "archetypal" myths and folktales provide, for more ancient or simpler cultures, prototypes for life narratives? And is it also the case that in more modern and more complex cultures, highly individualized life narratives parallel increased individualization both in the real life of individuals and in fictional narratives? Such questions are susceptible to adjudication by appeal to evidence. They offer rich opportunities for researchers capable of producing empirical data, and also for historical scholars capable of integrating data with scholarly information.

Ethologist Niko Tinbergen (1963) identifies four areas in which research into animal behavior should seek integrated answers: phylogeny, ontogeny, mechanism, and adaptive function. Phylogeny concerns the evolutionary history of a species and ontogeny the individual development of an organism within that species. Mechanisms consist of genetic, physiological, and neurological structures that produce the behavior. Hypotheses about adaptive function offer explanations for ways in which a behavior meets the needs of survival and reproduction. For the human proclivity to produce imaginative artifacts, a phylogenetic analysis would identify the way that proclivity evolved, the antecedent characteristics necessary for it to have evolved, when it first emerged as a distinct feature, and how it developed later in the evolutionary history of the species. In this volume, Wilson, Boehm, Rose, Dissanayake, and Carroll et al. give attention to that question. Analysis of the ontogeny of imaginative activity would focus on how and when that activity develops during childhood. Dissanayake and McAdams take up that issue, and Dissanayake develops an argument for parallels in the phylogeny and ontogeny of imaginative activity. Arguments about mechanisms of imaginative activity identify the cognitive, affective, and sensory equipment used for producing and consuming imaginative artifacts. Referencing research by Tomasello, Edelman, and others, Boyd discusses mechanisms of shared attention and pattern recognition. Clasen and Carroll et al. reference empirical research that identifies narrative as a form of "simulation." Hypotheses about the adaptive function of imaginative activity would explain how that activity contributes ultimately to the survival and reproduction of individuals living in groups and how it contributes to the success of groups competing with other groups.

Hypotheses about adaptive function have larger explanatory scope than hypotheses about the other three ethological questions. Understanding how a characteristic of an organism contributes to survival and reproduction provides crucial clues regarding why that characteristic has evolved in the way that it has, how it fits into the developmental trajectory of the animal, and how and why its mechanisms work as they do.

Contributors to this volume offer three main hypotheses, overlapping and complementary, about the adaptive function of the arts, especially narrative: building scenarios, internalizing social norms, and helping create a total imaginative universe. Wilson, McAdams, and Clasen all formulate versions of the idea that fictional narratives are scenarios that enable readers to envision alternative possible courses of action. McAdams and Carroll et al. postulate that narratives, autobiographical and fictional, help people internalize the beliefs and norms of their culture. This idea has a clear parallel with Boehm's arguments that humans evolved in such a way as to internalize cultural norms. Fictional narratives would be a cultural technology through which virtual or vicarious experience helps people to build morally valenced imaginative structures within which they can locate their own behavior. Dissanayake argues that all the arts are incorporated in rituals and ceremonies through which social groups affirm their collective identity and integrate individuals into the group. Though focusing on a highly particular subject, horror fiction, Clasen suggests the broadest encompassing theory about the adaptive function of the arts. He argues that horror fiction "gives us experience with negative emotion at levels of intensity not safely come by in real life. It thus allows us to incorporate the imagination of danger into our total imaginative universe."

The seminal formulation for the idea of a total imaginative universe appears in Wilson's *Consilience* (1998, Chapter 10; and see Carroll 2012). Wilson argues that cognitive and behavioral flexibility are defining characteristics of the human species. For animals of other species, the world presents itself as a series of stimuli or triggers that release a limited range of stereotyped responses. Humans, in contrast, build images of the world and regulate their behavior in accordance with those images. They locate the present in relation to a remembered past and a projected future; and they locate the visible world in relation to images and beliefs about unseen forces, some of them cosmic in scope. They envision their own behavior not just in terms of appetitive or aversive impulse, but also in terms of collective moral norms. In every individual mind, the image of the group, as a collective body, is an integral part of individual identity. One is a member of a tribe, a nation, an ethnic group, a political party, or a religious faith. Those groups are represented iconically and symbolically by decorative styles, personal ornamentation, and graphic designs—the cross, the hammer and sickle, the swastika, the national flag. Much earnest social idealism tries to think its way past these forms of group identity, but they are integral parts of our peculiarly human adaptive repertory. If we ever do manage to get past these particular forms of imagination, it will be only because we have succeeded in imagining that our individuality is intertwined with some still larger group—with humanity as a whole, or with the ecology of Earth as a whole.

A Direction for Future Research

Psychology tends to be ahistorical, working in synchronic schemas heavily oriented to contemporary populations, often a very small subset of contemporary populations. Evolutionary psychology occupies itself primarily with deep history, history on evolutionary timescales. Earlier phases of evolutionary psychology concentrated on the species-typical characteristics that were supposed to have become more or less fixed at some not very determinate point in the ancestral past (Barkow et al. 1992). Both biases—the fixation on the present and the fixation on a historically unvarying human nature—have illuminated important basic features of human behavior but have also slighted the emergent complexities of distinct cultures, distinct ecologies, and distinct historical periods. In contrast, cultural anthropologists, along with most contemporary literary scholars, have given an almost exclusive emphasis to specific cultures or culturally specific periods. For many of the historicists, there is no human nature, only distinct cultures. For many of the evolutionists, there has been no history, only a universal human nature. The next major phase in our understanding of human nature will come, I think, from scholars and scientists finding ways to investigate interactions between the biological and cultural parts of our nature. Fruitful research in that vein would combine ecological, socioeconomic, and political analysis of a particular society with an analysis of that society's imaginative culture, its ideologies, religions, philosophies, and literary and artistic practices. That kind of research would also incorporate general principles for large-scale social and political structures across historical time (Wilson 2002; Turchin 2006; Fukuyama 2011; Haidt 2012). And finally, ultimately, that kind of research would identify causal links between particular cultural ecologies and the basic elements of human nature—motives, emotions, features of personality, and forms of cognition (Chapter 12, this volume; Gottschall 2008). We are a biocultural species, but we have only just begun to conduct biocultural research. The essays in this volume offer strong indications for how that kind of research might be usefully developed.

References

Abrams, M. H. 1953. *The mirror and the lamp: Romantic theory and the critical tradition*. New York: Oxford Univerity Press.

———. 1965. Structure and style in the greater Romantic lyric. In Frederick W. Hilles and Harold Bloom, eds., *From sensibility to Romanticism: Essays presented to Frederick A. Pottle*, 527–560. Oxford University Press: New York.

———. 1997. The transformation of English studies: 1930–1995. *Dædalus: Journal of the American Academy of Arts and Sciences* 126: 105–131.

Alexander, Richard D. 1987. *The biology of moral systems*. Hawthorne, NY: de Gruyter.

Angier, Natalie. 2007. *The canon: A whirligig tour of the beautiful basics of science*. Boston: Houghton Mifflin.

Barkow, Jerome H., Leda Cosmides, and John Tooby, eds. 1992. *The adapted mind: Evolutionary psychology and the generation of culture*. New York: Oxford University Press.

Baron-Cohen, Simon. 2011. *The science of evil: On empathy and the origins of cruelty*. New York: Basic Books.

Baumeister, Roy F. 2005. *The cultural animal: Human nature, meaning, and social life*. Oxford: Oxford University Press.

Boehm, Christopher. 1999. *Hierarchy in the forest: The evolution of egalitarian behavior*. Cambridge, MA: Harvard University Press.

———. 2012. *Moral origins: The evolution of virtue, altruism, and shame*. New York: Basic Books.

Bourke, Andrew F. G. 2011. *Principles of social evolution*. Oxford: Oxford University Press.

Bowler, Peter J. 1988. *The non-Darwinian revolution: Reinterpreting a historical myth*. Baltimore: Johns Hopkins University Press.

Boyd, Brian. 2006. Getting it all wrong: Bioculture critiques cultural critique. *American Scholar* 75: 18–30.

———. 2009. *On the origin of stories: Evolution, cognition, and fiction*. Cambridge, MA: Harvard University Press.

Boyd, Brian, Joseph Carroll, and Jonathan Gottschall, eds. 2010. *Evolution, literature, and film: A reader*. New York: Columbia University Press.

Boyd, Robert, and Peter J. Richerson. 2007. Cultural adaptation and maladaptation: Of kayaks and commissars. In Steven W. Gangestad and Jeffry A. Simpson, eds., *The evolution of mind: Fundamental questions and controversies*, 327–331. New York: Guilford.

Boyd, Robert, and Joan B. Silk. 2012. *How humans evolved*. New York: Norton.

Brooks, David. 2011. *The social animal: The hidden sources of love, character, and achievement*. New York: Random House.

Brown, Donald E. 1991. *Human universals*. Philadelphia: Temple University Press.

Buss, David M., ed. 2005a. *The handbook of evolutionary psychology*. Hoboken, NJ: Wiley.

———. 2005b. *The murderer next door: Why the mind is designed to kill*. New York: Penguin.

Cacioppo, John T., and William Patrick. 2008. *Loneliness: Human nature and the need for social connection*. New York: Norton.

Carroll, Joseph. 1995. *Evolution and literary theory*. Columbia: University of Missouri Press.

———. 2001. The ecology of Victorian fiction. *Philosophy and Literature* 25: 295–313.

———. 2008a. An evolutionary paradigm for literary study. *Style* 42: 103–135.

———. 2008b. Rejoinder to the responses. *Style* 42: 308–411.

———. 2011a. Human life history and gene-culture co-evolution: An emerging paradigm. *Evolutionary Review: Art, Science, Culture* 2: 23–37.

———. 2011b. *Reading human nature: Literary Darwinism in theory and practice*. Albany: State University of New York Press.

———. 2012. The adaptive function of the arts: Alternative evolutionary hypotheses. In Carsten Gansel and Dirk Vanderbeke, eds., *Telling stories: Literature and evolution*, 50–63. Berlin: De Gruyter.

Carroll, Sean B. 2009. *Remarkable creatures: Epic adventures in the search for the origins of species*. Boston: Houghton Mifflin Harcourt.

Cochran, Gregory, and Henry Harpending. 2009. *The 10,000 year explosion: How civilization accelerated human evolution*. New York: Basic Books.

Damasio, Antonio R. 2010. *Self comes to mind: Constructing the conscious brain*. New York: Pantheon Books.

Darwin, Charles. 1859. *On the origin of species by means of natural selection*. London: J. Murray.

———. 1871. *The descent of man, and selection in relation to sex*. London: J. Murray.

Dawson, Gowan. 2006. Literature and science under the microscope. *Journal of Victorian Culture* 11: 301–315.

Degler, Carl N. 1991. *In search of human nature: The decline and revival of Darwinism in American social thought*. New York: Oxford University Press.

Deresiewicz, William. 2009. Adaptation: On literary Darwinism. *Nation*, June, 26–31.

Dissanayake, Ellen. 2000. *Art and intimacy: How the arts began*. Seattle: University of Washington Press.

———. 2011. In the beginning, evolution created religion and the arts. *Evolutionary Review: Art, Science, Culture* 2: 64–81.

Dunbar, R. I. M., and Louise Barrett, eds. 2007. *Oxford handbook of evolutionary psychology*. Oxford: Oxford University Press.

Dutton, Denis. 2009. *The art instinct: Beauty, pleasure, and human evolution*. New York: Bloomsbury Press.

Easterlin, Nancy. 2004. "Loving ourselves best of all": Ecocriticism and the adapted mind. *Mosaic* 37: 1–18.

Eibl-Eibesfeldt, Irenäus. 1989. *Human ethology*. New York: Aldine De Gruyter.

Flatt, Thomas, and Andreas Heyland, eds. 2011. *Mechanisms of life history evolution: The genetics and physiology of life history traits and trade-offs*. Oxford: Oxford University Press.

Foley, Robert. 1995. The adaptive legacy of human evolution: A search for the environment of evolutionary adaptedness. *Evolutionary Anthropology: Issues, News, and Reviews* 4: 194–203.

Fox, Robin. 1989. *The search for society: Quest for a biosocial science and morality*. New Brunswick, NJ: Rutgers University Press.

Freeman, Derek. 1983. *Margaret Mead and Samoa: The making and unmaking of an anthropological myth*. Cambridge, MA: Harvard University Press.

Freud, Sigmund. (1930/1961). *Civilization and its discontents*. In James Strachey, ed., *The standard edition of the complete psychological works of Sigmund Freud*, vol. 21. London: Hogarth.

Fukuyama, Francis. 2011. *The origins of political order: From prehuman times to the French Revolution*. New York: Farrar, Straus and Giroux.

Gangestad, Steven W., and Jeffry A. Simpson. 2007. *The evolution of mind: Fundamental questions and controversies*. New York: Guilford Press.

Geary, David C. 2005. *The origin of mind: Evolution of brain, cognition, and general intelligence*. Washington, DC: American Psychological Association.

Geary, David C., and Mark V. Flinn. 2001. Evolution of human parental behavior and the human family. *Parenting: Science and Practice* 1: 5–61.

Goleman, Daniel. 2006. *Social intelligence: The new science of human relationships*. New York: Bantam.

Goodheart, Eugene. 2007. *Darwinian misadventures in the humanities*. New Brunswick, NJ: Transaction.

Gottschall, Jonathan. 2008. *The rape of Troy: Evolution, violence, and the world of Homer*. Cambridge: Cambridge University Press.

———. 2012. *The storytelling animal: How stories make us human*. Boston: Houghton Mifflin Harcourt.

Haidt, Jonathan. 2006. *The happiness hypothesis: Finding modern truth in ancient wisdom*. New York: Basic Books.

———. 2012. *The righteous mind: Why good people are divided by politics and religion*. New York: Pantheon Books.

Hamilton, William D. 1964a. The genetical evolution of social behaviour, I. *Journal of Theoretical Biology* 7: 1–16.

———. 1964b. The genetical evolution of social behaviour, II. *Journal of Theoretical Biology* 7: 17–52.

Hawkins, Jeff, with Sandra Blakeslee. 2004. *On intelligence*. New York: Times Books.

Huxley, Julian. 1942. *Evolution: The modern synthesis*. London: G. Allen & Unwin.

Huxley, Thomas Henry. 1863. *Evidence as to man's place in nature*. London: Williams & Norgate.

Irons, William. 1998. Adaptively relevant environments versus the environment of evolutionary adaptedness. *Evolutionary Anthropology* 6: 194–204.

Kaplan, Hillard S., and Steven W. Gangestad. 2005. Life history theory and evolutionary psychology. In David M. Buss, ed., *The handbook of evolutionary psychology*, 68–95. Hoboken, NJ: Wiley.

Kean, Sam. 2011. *The disappearing spoon: And other true tales of madness, love, and the history of the world from the periodic table of the elements*. New York: Back Bay Books.

———. 2012. *The violinist's thumb: And other lost tales of love, war, and genius, as written by our genetic code*. New York: Little, Brown.

Kenrick, Douglas T. 2011. *Sex, murder, and the meaning of life: A psychologist investigates how evolution, cognition, and complexity are revolutionizing our view of human nature*. New York: Basic Books.

Klein, Richard G. 2009. *The human career: Human biological and cultural origins*. Chicago: The University of Chicago Press.

Klein, Richard G., with Blake Edgar. 2002. *The dawn of human culture*. New York: Wiley.

Kramnick, Jonathan. 2011. Against literary Darwinism. *Critical Inquiry* 37: 315–347.

Laland, Kevin N., and Gillian R. Brown. 2011. *Sense and nonsense: Evolutionary perspectives on human behaviour*. New York: Oxford University Press.

Lane, Nick. 2009. *Life ascending: The ten great inventions of evolution*. New York: Norton.

Linden, David J. 2007. *The accidental mind*. Cambridge, MA: Harvard University Press.

———. 2011. *The compass of pleasure: How our brains make fatty foods, orgasm, exercise, marijuana, generosity, vodka, learning, and gambling feel so good*. New York: Viking.

Lummaa, Virpi. 2007. Life history theory, reproduction and longevity in humans. In R. I. M. Dunbar and Louise Barrett, eds., *Oxford handbook of evolutionary psychology*, 397–414. Oxford: Oxford University Press.

Maynard Smith, John, and Eörs Szathmáry. 1995. *The major transitions in evolution*. Oxford: W.H. Freeman Spektrum.

Mayr, Ernst. 1982. *The growth of biological thought: Diversity, evolution, and inheritance*. Cambridge, MA: Harvard University Press.

McAdams, Dan P. 2006. *The redemptive self: Stories Americans live by*. Oxford: Oxford University Press.

McKeon, Michael. 1987. *The origins of the English novel, 1600–1740*. Baltimore: Johns Hopkins University Press.

Mellars, Paul, ed. 2007. *Rethinking the human revolution: New behavioural and biological perspectives on the origin and dispersal of modern humans*. Cambridge, UK: McDonald Institute for Archaeological Research.

Mellars, Paul, and Christopher B. Stringer, eds. 1989. *The human revolution: Behavioural and biological perspectives on the origins of modern humans*. Princeton, NJ: Princeton University Press.

Miller, Geoffrey F. 2000. *The mating mind: How sexual choice shaped the evolution of human nature*. New York: Doubleday.

Mithen, Steven J. 1996. *The prehistory of the mind: A search for the origins of art, religion, and science*. London: Thames and Hudson.

———. 2004. *After the ice: A global human history, 20,000–5000 BC*. Cambridge, MA: Harvard University Press.

Nettle, Daniel. 2006. The evolution of personality variation in humans and other animals. *American Psychologist* 61: 622–631.

———. 2007. Individual differences. In Robin I. M. Dunbar and Louise Barrett, eds., *Oxford handbook of evolutionary psychology*, 479–490. Oxford: Oxford University Press.

Oakley, Barbara A. 2007. *Evil genes: Why Rome fell, Hitler rose, Enron failed and my sister stole my mother's boyfriend*. Amherst, NY: Prometheus Books.

Panksepp, Jaak, and Lucy Biven. 2012. *The archaeology of mind: Neuroevolutionary origins of human emotions*. New York: Norton.

Pinker, Steven. 1997. *How the mind works*. New York: Norton.

———. 2002. *The blank slate: The modern denial of human nature*. New York: Viking.

———. 2005. Foreword. In David M. Buss, ed., *The handbook of evolutionary psychology*, xi–xvi. Hoboken, NJ: Wiley.

———. 2011. *The better angels of our nature: Why violence has declined*. New York: Viking.

Potts, Rick. 1998. Variability selection in hominid evolution. *Evolutionary Anthropology* 7: 81–96.

Pross, Addy. 2012. *What is life? How chemistry becomes biology*. Oxford: Oxford University Press.

Richerson, Peter J., and Robert Boyd. 2005. *Not by genes alone: How culture transformed human evolution*. Chicago: University of Chicago Press.

Ridley, Matt. 1994. *The Red Queen: Sex and the evolution of human nature*. New York: Macmillan.

———. 1996. *The origins of virtue*. London: Viking.

Shackelford, Todd K., and Viviana A. Weekes-Shackelford. 2012. *The Oxford handbook of evolutionary perspectives on violence, homicide, and war*. Oxford: Oxford University Press.

Shubin, Neil. 2008. *Your inner fish: A journey into the 3.5-billion-year history of the human body*. New York: Pantheon.

———. 2013. *The universe within: Discovering the common history of rocks, planets, and people*. New York: Pantheon Books.

Smith, Eric A, Monique Borgerhoff Mulder, and Kim Hill. 2001. Controversies in the evolutionary social sciences: A guide for the perplexed. *Trends in Ecology & Evolution* 16: 128–135.

Sober, Elliott, and David Sloan Wilson. 1998. *Unto others: The evolution and psychology of unselfish behavior*. Cambridge, MA: Harvard University Press.

Sterelny, Kim. 2003. *Thought in a hostile world: The evolution of human cognition*. Malden, MA: Blackwell.

Symons, Donald. 1992. On the use and misuse of Darwinism in the study of human behavior. In Jerome H. Barkow, Leda Cosmides, and John Tooby, eds., *The adapted mind: Evolutionary psychology and the generation of culture.*, 137–159. New York: Oxford University Press.

Thagard, Paul. 2010. *The brain and the meaning of life*. Princeton, NJ: Princeton University Press.

Tinbergen, Niko. 1963. On aims and methods of ethology. *Zeitschrift für Tierpsychologie* 20: 410–433.

Tomasello, Michael, Malinda Carpenter, Josep Call, Tanya Behne, and Henrike Moll. 2005. Understanding and sharing intentions: The origins of cultural cognition. *Behavioral and Brain Sciences* 28: 675–691.

Tooby, John, and Leda Cosmides. 1992. The psychological foundations of culture. In Jerome H. Barkow, Leda Cosmides, and John Tooby, eds., *The adapted mind: Evolutionary psychology and the generation of culture.*, 19–136. New York: Oxford University Press.

Trivers, Robert L. 1971. The evolution of reciprocal altruism. *Quarterly Review of Biology* 46: 35–57.

———. 1972. Parental investment and sexual selection. In Bernard Campbell, ed., *Sexual selection and the descent of man*, 136–179. Chicago: Aldine.

Turchin, Peter. 2006. *War and peace and war: The life cycles of imperial nations*. New York: Pi.

Wade, Nicholas. 2006. *Before the dawn: Recovering the lost history of our ancestors*. New York: Penguin.

Watt, Ian P. 1957. *The rise of the novel: Studies in Defoe, Richardson, and Fielding.* Berkeley: University of California Press.

Weinberg, Steven. 1992. *Dreams of a final theory.* New York: Pantheon Books.

———. 1993. *The first three minutes: A modern view of the origin of the universe.* New York: Basic Books.

Wellek, René. 1949a. The concept of Romanticism in literary history I: The term Romantic and its derivatives. *Comparative Literature* 1: 1–23.

———. 1949b. The concept of Romanticism in literary history II: The unity of European Romanticism. *Comparative Literature* 1: 147–172.

———. 1961. The concept of realism in literary scholarship. *Neophilologus* 45: 1–20.

Wilson, David Sloan. 2002. *Darwin's cathedral: Evolution, religion, and the nature of society.* Chicago: University of Chicago Press.

———. 2007. Group-level evolutionary processes. In Robin I. M. Dunbar and Louise Barrett, eds., *Oxford handbook of evolutionary psychology*, 49–55. Oxford: Oxford University Press.

Wilson, Edmund. 1931. Symbolism. Chapter one in *Axel's castle: A study in the imaginative literature of 1870–1930*, 1–25. New York: Charles Scribner's & Sons.

Wilson, Edward O. 1992. *The diversity of life.* Cambridge, MA: Harvard University Press.

———. 1998. *Consilience: The unity of knowledge.* New York: Knopf.

———. 2012. *The social conquest of earth.* New York: Liveright.

Wrangham, Richard W. 2009. *Catching fire: How cooking made us human.* New York: Basic Books.

Zuk, Marlene. 2013. *Paleofantasy: What evolution really tells us about sex, diet, and how we live.* New York: Norton.

I TRANSFORMING OUR VISION OF THE HUMAN STORY

1 THE MEANING OF HUMAN EXISTENCE

Edward O. Wilson

The meaning of humanity is too important a subject to leave to the humanities. Their many branches—from philosophy to law, from history to the creative arts—have described the particularities of human nature with genius and exquisite detail, back and forth in endless permutations. But, they have not explained *why* we possess our special nature and not some other out of a vast number conceivable. In that sense, the humanities have not accounted for the meaning of our species' existence.

So, just what are we? The key to the great riddle lies in the circumstance and process that created our species. The human condition is a product of history, not just the six millennia of civilization, but very much further back, across hundreds of millennia. The whole of it—biological and cultural evolution, in seamless unity—must be explored for an answer to the mystery. When thus viewed across its entire traverse, the history of humanity also becomes the meaning of the species.

A majority of people prefer to interpret history as the unfolding of a supernatural design, to whose author we owe obedience. But that comforting interpretation has grown less supportable as knowledge of the real world has expanded. Scientific knowledge in particular has been doubling every 10 to 20 years for more than a century. In traditional explanations of the past, religious creation stories have been blended with the humanities to attribute meaning to our species. It is time to consider what science might give to the humanities and the humanities to science in a common search for a more solidly grounded answer to the great riddle.

To begin, biologists have found that the biological origin of advanced social behavior in humans was similar to that occurring elsewhere in the Animal Kingdom. Using comparative studies of thousands of animal species, from insects to mammals, they have concluded that the most complex societies have arisen through eusociality—roughly, "true" social condition. The members of a eusocial group cooperatively rear the young across multiple generations. They also divide labor through the surrender by some members of at least some of their personal reproduction in a way that increases the personal reproduction of other members.

Eusociality stands out as an oddity in a couple of ways. One is its extreme rarity. Out of hundreds of thousands of evolving lines of animals on the land during the past 400 million years, the condition, so far as we can determine, has arisen only about two dozen times. This is likely to be an underestimate as a result of sampling error. Nevertheless, we can be certain that the number of originations was very small.

Furthermore, the known eusocial species arose very late in the history of life. It appears to have occurred not at all during the great Paleozoic diversification of insects, 350 million years to 250 million years before the present, during which the variety of insects approached that of today. Nor is there, as yet, any evidence of eusocial species during the Mesozoic Era until the appearance of the earliest termites and ants between 200 million years ago and 150 million years ago. Humans at the *Homo* level appeared only very recently, following tens of millions of years of evolution among the primates.

Once attained, advanced social behavior at the eusocial grade has proved a major ecological success. Of the two dozen independent lines, just two within the insects, ants and termites, globally dominate invertebrates on the land. Although they are represented by fewer than 20,000 of the million known living insect species, ants and termites compose more than one-half of insect body weight.

The history of eusociality creates a dilemma: given the enormous advantage it confers, why was this advanced form of social behavior so rare and long delayed? The answer appears to be the special sequence of preliminary evolutionary changes that must occur before the final step to eusociality can be taken. In all the eusocial species analyzed to date, the final step before eusociality is the construction of a protected nest, from which foraging trips are launched and within which the young are raised to maturity. The original nest builders can be a lone female, a mated pair, or a small and weakly organized group. When this final preliminary step is attained, all that is needed to create a eusocial colony is for the parents and offspring to stay at the nest and cooperate in raising additional generations of young. Such primitive assemblages then divide easily into risk-prone foragers and risk-averse parents and nurses.

What brought one primate line to the rare level of eusociality? Paleontologists have found that the circumstances were humble. In Africa, about two million years ago, one species of the primarily vegetarian australopithecine evidently shifted its diet to include a much greater reliance on meat. For a group to harvest such a high-energy, widely dispersed source of food, it did not pay to roam about as a loosely organized pack of adults and young like present-day chimpanzees and bonobos. It was more efficient to occupy a campsite (thus, the nest) and send hunters out who bring home meat either killed or scavenged to share with others. In exchange, the hunters received protection of the campsite and their own young offspring kept there.

From studies of modern humans, including hunter-gatherers, whose lives tell us so much about human origins, social psychologists have deduced the mental growth that began with hunting and campsites. A premium was placed on personal relationships geared to both competition and cooperation among the members. The process was ceaselessly dynamic and demanding. It far exceeded in intensity anything similar experienced by the roaming, loosely organized bands of most animal societies. It required a memory good enough to assess the intentions of fellow members, to predict their responses from one moment to the next, and it resulted in the ability to invent and inwardly rehearse competing scenarios of future interactions.

The social intelligence of the campsite-anchored prehumans evolved as a kind of nonstop game of chess. Today, at the terminus of this evolutionary process, our immense memory banks are smoothly activated across the past, present, and future. They allow us to evaluate the prospects and consequences variously of alliances, bonding, sexual contact, rivalries, domination, deception, loyalty, and betrayal. We instinctively delight in the telling of countless stories about others as players on the inner stage. The best of it is expressed in the creative arts, political theory, and other higher level activities we have come to call the humanities.

The definitive part of the long creation story evidently began with the primitive *Homo habilis* (or a species closely related to it) two million years ago. Before the habilines, the prehumans had been animals. Largely vegetarians, they had humanlike bodies, but their cranial capacity remained chimpanzee-size, at or below 500 cc. Starting with the habiline period, the capacity grew precipitously to 680 cc in *Homo habilis*, 900 cc in *Homo erectus*, and about 1400 cc in *Homo sapiens*. The expansion of the human brain was one of the most rapid episodes of evolution of complex organisms in the history of life.

Still, to recognize the rare coming together of cooperating primates is not enough to account for the full potential of modern humans that brain capacity provides. Evolutionary biologists have searched for the grand master of advanced social evolution, the combination of forces and environmental circumstances that bestowed greater longevity and more successful reproduction on the possession of high social intelligence. Currently, there are two competing theories of the principal force. The first is kin selection: individuals favor collateral kin (relatives other than offspring), making it easier for altruism to evolve among members of the same group. Altruism, in turn, engenders complex social organization and, in the one case that involves big mammals, human-level intelligence.

The second, more recently argued theory (full disclosure: I am one of the modern version's authors), the grand master, is multilevel selection. This formulation recognizes two levels at which natural selection operates: individual selection based on competition and cooperation among members of the same group, and group selection, which arises from competition and cooperation between groups. Multilevel selection is gaining in favor among evolutionary biologists because of a recent mathematical proof that kin selection can arise only under special conditions that demonstrably do not exist, and the better fit of multilevel selection to all of the two dozen known animal cases of eusocial evolution.

The roles of both individual and group selection are indelibly stamped (to borrow a phrase from Charles Darwin) on our social behavior. As expected, we are intensely interested in the minutiae of behavior of those around us. Gossip is a prevailing subject of conversation, everywhere from hunter-gatherer campsites to royal courts. The mind is a kaleidoscopically shifting map of others, each of whom are drawn emotionally in shades of trust, love, hatred, suspicion, admiration, envy, and sociability. We are compulsively driven to create and belong to groups, variously nested, overlapping, or separate, and large or small. Almost all groups compete with those of similar kind in some manner or the other. We tend to think of our own as superior, and we find our identity within them.

The existence of competition and conflict, the latter often violent, has been a hallmark of societies as far back as archeological evidence is able to offer. These and other traits we call human nature are so deeply resident in our emotions and habits of thought as to seem just part of some greater nature, like the air we all breathe, and the molecular machinery that drives all of life. But they are not. Instead, they are among the idiosyncratic hereditary traits that define our species.

The major features of the biological origins of our species are coming into focus, and with this clarification comes the potential of a more fruitful contact between science and the humanities. The convergence between these two great branches of learning will matter hugely when enough people have thought it through. On the science side, genetics, the brain sciences, evolutionary biology, and paleontology will be seen in a different light. Students will be taught prehistory as well as conventional history, the whole presented as the living world's greatest epic.

We will also, I believe, take a more serious look at our place in nature. Exalted we are, indeed, risen to be the mind of the biosphere without a doubt, our spirits capable of awe and ever-more breathtaking leaps of imagination. But we are still part of Earth's fauna and flora. We are bound to it by emotion, physiology, and, not least, deep history. It is dangerous to think of this planet as a way station to a better world or to continue to convert it into a literal, human-engineered spaceship. Contrary to general opinion, demons and gods do not vie for our allegiance. We are self-made, independent, alone, and fragile. Self-understanding is what counts for long-term survival, both for individuals and for the species.

Author's Note

This essay was modified from an online article in the *New York Times* Opinionator Section, February 24, 2013, under the title "The Riddle of the Human Species."

II THE EVOLUTION OF HUMAN SOCIALITY

2 BULLIES

REDEFINING THE HUMAN FREE-RIDER PROBLEM

Christopher Boehm

When Charles Darwin (1871) added group selection to the basically "individualistic" paradigm he created, he briefly considered the problem of free-riders, and did so mainly by focusing on cowards riding on the backs of heroes. However, Darwin lacked both the knowledge of genes and the mathematical modeling that have made the new synthesis in biology so powerful. When Mendelian units and mathematical models finally joined with natural selection theory, new possibilities arose.

The biologist who reintroduced the free-rider problem in this far more sophisticated context was George Williams (1966), who did so by attacking poorly documented theories that assumed either group selection operating within species, or else species selection operating as a larger version of the same model. The ethological work of Wynne-Edwards (1962) with birds that stopped reproducing when food supplies became scarce was the main target, but Lorenz's (1966) species selection assumptions also became seriously at risk. Williams's attack was, perhaps, too successful—as things turned out.

As a result, Darwin's hypothesis that group selection could be a significant evolutionary force in social species was sent to the dog house, and, for the great majority of evolutionary biologists, after Edward O. Wilson's (1975b) agreement with Williams's assessment, it stayed in the dog house for more than three decades. Beginning with D. S. Wilson's early crusading work (Wilson 1975a), however, a few dissidents like myself (Boehm 1978, 1997) refused to concede the point and group selection theory has now achieved a respectable comeback (Wilson and Wilson 2007) as part of multilevel selection theory. Despite this, however, maladapatively generous behavior in altruistic humans still remains to be fully explained.

On the basis of group selection's mechanical weakness in comparison with selection taking place within groups, Williams and his many followers attacked group selection theory to the point that it became a pariah paradigm, whereas another effect of his work was to stimulate a long-lasting focus on free-riding cheaters. This was because they posed a serious obstacle to the genetic selection of altruistic traits in species such as humans, and these cheaters gained further renown at the hands of Robert Trivers (1971). Trivers discussed their predatory cheating behavior as an obstacle to reciprocal altruism—a necessarily tit-for-tat process that involved mutual aid being exchanged between two cooperating partners over time, and the result was that collaborators would outcompete others who failed to cooperate, as long as neither partner took a free ride.

Trivers identified such relationships, along with cheating behaviors, in a number of species, and with group selection basically being "off limits" for decades, a cheater-free reciprocal-altruism paradigm became a major explanation for humans' self-sacrificial helping of a nonkinsman. In fact, along with kin selection (Hamilton 1964), reciprocal altruism became the paradigm of choice, with the latter being the model that could explain cooperation that went beyond nepotism. Because cheaters could undermine the selection process that supported such generosity, Cosmides et al. (2005) suggested that there could be a dedicated, evolved cheater detection mechanism that protected these "altruists" because, once identified, such deceivers could be avoided.

As social predators, in effect free-riders were defined as deceivers who were dedicated to taking advantage of gullible, generous altruists. The immediate strategies of these predators were consistent with the many references to free-riding in economics and public policy, where just at the level of phenotype cheating plays a major role in creating practical problems for social cooperation (Hardin 1968). However, they were now being defined as important "players" in the formation and maintenance of human gene pools, and also as key players with respect to altruism's possibilities.

Enter the Bully as Potent Free-Rider

Free-riding can be defined as a relationship in which one person gains reproductively at the expense of another by deceptively taking advantage of a situation in which the other party is vulnerable because of being generous. However, there is another type of free-riding in which the potential advantages may be still greater than those obtainable by cheaters. Here I have in mind the *dominant* free-rider, a bully who takes reproductive advantage by means of force or threat of force, rather than by deceit (Boehm 1997).

Although the fitness advantages of bullying are far from being unknown, such behavior has been little discussed in the context of altruism and its selection. A major difference between a free-riding cheater and a free-riding bully is that a cheater evolves as a dedicated selfish predator on altruists, whereas a bully evolves as a far more generalized selfish predator who preys on anyone who ranks lower in a social dominance hierarchy (Boehm 1999). This means that a bully can take advantage of altruists and nonaltruists alike, as long as they are lower ranking, less powerful, or less motivated to dominate. This means that when two equally strong individuals are vying for dominance and one of them is an altruist and the other isn't, the nonaltruist will be selfishly motivated to gain the edge and take advantage of his altruistic competitor.

Although the general reproductive advantages obtained by dominators can be assessed by common sense, eventually such assumptions were tested by Ellis (1995) for a variety of mammals including primates, whereas for culturally hierarchical humans, Betzig (1986) showed that alphas were gaining an edge that could be very substantial, indeed.

Social Dominance

In my view, social dominance hierarchies are based on three factors: social competition, dominance, and submission (Boehm 1999). Not all mammalian species have such hierarchies, but many do, including social carnivores such as lions, wolves, or dolphins whose pecking orders regulate the sharing of meat; omnivorous primates such as chimpanzees, bonobos, capuchin monkeys, and humans; and plant-eaters like gorillas and small-brained chickens. It was in studying captive chickens that Schjelderup-Ebbe (1935) first documented nonlinear pecking orders, but some of these species have highly specified linear orders in which every individual falls into a consistent system of descending rank that applies to every individual involved. It is the dominators, obviously, who are the selfish bullies; their victims are the submitters whose ranks, as seen earlier, will include disproportionate numbers of altruists.

These ranked societies have alpha males as ultimate bullies, but there has been a partial exception. Humans have a long evolutionary history of being egalitarian, and therefore without alpha males, although after the rise of agriculture our species suddenly became prone to hierarchy (Knauft 1991). For those earlier humans, there were only two ways to escape being dominated: get rid of the dominators (Boehm 1993) or exit the group (Vehrencamp 1983) to seek one with less dominant competitors.

Thus, in evolutionary studies of human altruism, the *dominant free-rider* must be seen as a major problem, and one that is separate from the issue of cheating free-riders. Here I review a number of theoretical models (Boehm 2012b) that have been proposed for the genetic explanation of altruistic behavior, and consider the possibilities for both of these free-rider types, before an analysis of bullying free-rider suppression in humans is undertaken.

Relevant Models for Analysis of Altruism

1. *Basic selection mechanisms at the level of individuals, including inclusive fitness.* Darwinian selection involves individuals as units of heredity and should disfavor altruists because free-riders will take advantage of them. Of course, if I help my close relative at a modest cost to myself but with substantial benefit to this kinsman, in effect I am helping my own genes (Hamilton 1964). Technically, this should be termed *nepotism* rather than *altruism*, even though genetically self-sacrificial generosity is involved (Campbell 1975).
2. *Group selection* (Wilson and Wilson 2007) can become significant if group extinction rates and phenotypic variation *between* groups become strong enough to counter the effects of similar forces that act *within* groups. Both dominant free-riding and cheating free-riding can undermine such effects insofar as altruists can be taken advantage of by deceivers or bullies, and this will drive down the frequencies of their genes even as group selection is trying to support them.
3. *Reciprocal altruism* (Trivers 1971) can become significant if pairs of individuals who cooperate can outcompete pairs who do not; however, either cheaters or bullies can take advantage of their partners and thereby undermine the basis of the cooperative process by contributing less than they receive.
4. *Mutualism* (West et al. 2007) involves a far more immediate version of reciprocal altruism, and in considering one-shot cooperation episodes cheating is considered to be a nonfactor because of the cooperation's immediacy; however, once the cooperation has yielded results, a bully could appropriate a disproportionate share of the mutualistic benefits, and thereby undermine the process genetically.
5. *Indirect reciprocity* (Alexander 1987) is involved when I help another in need without any payback, and then a third party helps me when I am in need. The selection basis for such altruism is found in reputational benefits: helpers with good reputations are favored in marriage choices and in the selection of other types of economic partners, and thereby any fitness losses resulting from altruistic acts are more than compensated. This theory encompasses costly signaling (Zahavi 1995; Zahavi et al. 1997; Pilot 2005), a model that has enjoyed an overworked popularity similar to that of reciprocal altruism. Basically, costly signaling is just another version of Alexander's theory, which D. S. Wilson (1999) calls "selection-by-reputation." A cheater might

dissemble being altruistic and therefore socially desirable, but in humans this would likely be detectible. A bully might try to go against such a system simply by using his power to take what he wanted, rather than waiting to be chosen, but in doing so he would suffer a bad reputation with its reproductive consequences and therefore might come out behind (Boehm 1997).

6. *Social selection* (West-Eberhard 1979; Nesse 2007; Boehm 2008a) is something of a catch-all category, which by my inclusive definition for humans includes selection-by-reputation, treated earlier, and also preaching in favor of altruism and sanctioning according to group social preferences that disfavor social predation, along with any other kind of selection process in which a larger variety of socially "impactful" personal preferences (West-Eberhard 1975) may affect genetic outcomes. Social selection acts against both bullies and cheaters.
7. *Docility-based piggybacking effects* come into play because humans are evolved to absorb their local culture automatically (Simon 1990). The reproductive benefits of not having to learn such useful behaviors by costly trial and error are large, but a secondary effect is that individuals will tend to conform to cultural messages that favor altruism, which are universal and also prominent among nomadic foragers (Boehm 2008b). With this model, altruism can be considered a pleiotropic or "piggybacking" effect (Gintis 2003) insofar as a moderately costly altruistic trait can be supported by the very substantial general benefits of cultural docility. Free-riders are not relevant to this model.
8. *Misplaced nepotism* (Boehm 1981) can take place when benefits normally conferred on close kin are also conferred on nonkin because kin recognition mechanisms are less than perfect in their screening functions. The fact that hunter-gatherers often call nonkin they are especially bonded to by terms that technically refer to *kinsmen* suggests this takes place frequently, and this, too, is a piggybacking model in that the great inclusive fitness benefits that come from helping close kin are "subsidizing" the losses incurred when closely bonded nonkin are helped in similar ways. A cheating nonkin free-rider might be able to behave in kinlike ways, and receive such benefits.

In the first five models, altruists are vulnerable to bullying behavior and often to cheating behavior. This would suggest that altruism is even less likely to evolve genetically than it appears when only cheating free-riders are considered. However, with humans, and to a lesser degree with the two *Pan* species, there are social countermeasures that are likely to have ameliorated or negated the effects of dominant free-riding by bullies. Here, the focus is on these power-based free-riders because, unlike cheaters, they have not been a subject of intensive analysis for the past half century.

The Natural History of Human Bullies

Ancestral *Pan* is the shared ancestor of humans, bonobos, and chimpanzees (Wrangham 1987; Ruvolo et al. 1991; Lion et al. 2011), and any behavior identified in all three of the extant species may be assumed, on the basis of parsimony (Boehm 2012a), to have been present in this ancestor. Chimpanzees and bonobos have clearly marked social dominance hierarchies with alpha males and high-ranking or alpha females, whereas for nomadic human hunter-gatherers, who are the appropriate group for evolutionary analysis of this type, hierarchical tendencies are strongly present but the hierarchies are *inverted* (Boehm 1984) with the subordinates definitively suppressing alpha-type behavior. The result is a "reverse dominance hierarchy" (Boehm 1993).

Chimpanzees and bonobos also suppress alpha behavior, but far less definitively (Boehm 1999), and in all three species this takes place by subordinates forming political coalitions (Harcourt and de Waal 1992) that enable them to go up against the bullies in their midst. In assessing the ancestral pattern it makes sense to operate conservatively, using a least-common-denominator principle. This means the ancestral level of bully suppression would need to be assessed at the level of either chimpanzees or bonobos, whichever species does the least in this area, because among hunter-gatherers the suppression of dominant behavior is far more definitive.

What is clear, is that neither *Pan* species approaches the effects achieved by egalitarian humans, who curb alpha-male behavior so effectively and so preemptively that even the politically strongest males behave with humility and don't try to boss around or otherwise dominate or despoil other adult hunters. In the case of wild bonobos, females regularly form dyadic alliances to compete with males in feeding (de Waal and Lanting 1997), and once in a while serious domination can be countered by a large group of females and males attacking a bully and seriously wounding him (Parker 2007; Hohmann and Fruth 2011). Among wild chimpanzees, dyadic alliances of males compete with the alpha and his allies in an attempt to unseat and replace him, and their efforts may ameliorate his power. In addition, in the wild, occasionally large groups may attack disliked high-ranking bullies and wound, exile, or kill them (Goodall 1992; Nishida 1996). Furthermore, among captive chimpanzees, the females form strong coalitions that prevent males from redirecting aggression at females, and often they are in a position to choose the alpha male who will be mediating their conflicts (de Waal 1982, 1996).

Thus, Ancestral *Pan* had dominance and submission tendencies that made for social dominance hierarchies, but this ancestor also disliked being dominated and was able to form coalitions that at least in part reduced the levels of domination by high-ranking group members. This suppression of bullying would at least have reduced the capacity of dominants to take reproductive advantage of otherwise evenly matched selfish altruists or others who were less dominant. In contrast, there is no good evidence from all three of these species that would suggest the presence of active punishment of cheaters, even though individual "defections" are reported for humans and also in the case of dyadic partnerships of chimpanzees that involve political alliancing (de Waal 1982, 1996) or meat-sharing (Watts and Mitani 2002; Mitani 2005).

Social Control in Earlier Humans

Consider where human evolution took us, and the end product is anatomically and culturally modern hunter-gatherers. Our species reached this advanced status 45,000 years ago (Klein 1999), and possibly earlier (McBrearty and Brooks 2000). Living in small bands composed mainly of nonkin (Hill et al. 2011), such people may be assumed to be essentially very similar to today's nomadic foragers: they would have been politically egalitarian, moralistic, and intent on achieving social control over social predators (Boehm 2000), and in every other way very similar. For that reason, universal or highly predominant behavior patterns among today's independent nomadic foragers can be assumed in these forbears.

To keep prehistoric reconstructions appropriate to the Late Pleistocene, the methodology I have used is as follows. I began with Binford's (2001) list of 339 purely foraging societies, and triaged it by removing (a) societies involved with domestication such as mounted hunters or those that practice some horticulture, (b) hunters involved in the fur trade, and (c) foragers who had become both sedentary and hierarchical. This left approximately half the list—societies that were economically independent, mobile, and, politically speaking, predictably egalitarian. As a type, I have called them "Late-Pleistocene appropriate" (LPA), and I have analyzed statistically the social control patterns of 49 of these societies with respect to bullying and cheating behaviors (Boehm 2012b). Here I summarize some relevant findings.

Capital punishment is reported for about half the 49 LPA foragers I assessed, and in all probability there has been some ethnographic underassessment because people such as !Kung Bushmen quickly learn that colonial or other authorities will punish indigenous capital punishment although, in the indigenous view, this is seen as being moral (Lee 1979). It is assumed that with much larger ethnographic time samples, and without this indigenous tendency to suppress information, reports of capital punishment would include many more societies in this sample. Thus, I suggest that 45,000 years ago capital punishment was very widespread and possibly even universal, and that although the behavior probably was quite rare in most of those societies, its reproductive consequences were powerful.

Did such ultimate punishment apply more to bullies or to cheaters? With respect to the offenses, half of the 49 contemporary ethnographically studied societies reported bullying behavior as a cause for killing a deviant, whereas sexual offenses were far less frequently cited, and cheating, in the form of theft or other deception, was reported only twice (Boehm 2012b). In terms of consequences, we may roughly assume that most of the deviants were male age 20 to 40 years, which would make the reproductive effects of lethal sanctioning strong because, on average, a significant portion of their reproductive career would have been curtailed.

If we look at what happens with individuals especially prone to act the bully in these 49 LPA societies, many of them are not severely punished for two reasons. One is that they are cowed by group sanctions, or the threat thereof, to a degree that they refrain from dominating others despite their tendencies to do so; the same goes for cheaters with respect to theft or deception. Second, hunter-gatherers understand very well the functions of gossiping in their small societies, for they all participate in gossip networks and they know that just as they exchange information about mainly bad but also good behaviors of others, they themselves will also be the subjects of gossip. This not only ensures detection, but also it raises the possibility of collective punishment because gossip can lead to a group consensus.

Thus, gossip permits group opinion to build through exchange of information and judgment, even when private gossiping sessions remain merely dyadic. Gossiping is done in a situation of mutual trust, and in this way public opinion can form safely and reach a consensus even with respect to a serious and dangerous deviant. In this way, a decision to "off" a serious deviant, usually a bully, can be made with minimal risk to fearful, exploited, or outraged decision makers.

Once the decision is made, there are practical problems faced in removing a serious and unreformable deviant by killing him; it's almost always a male. The close kin of such a deviant may realize he is a threat to other people or to the entire social order, but because of social bonding at the family level they will, in their grief and anger, be prone to retaliate lethally (Boehm 2011). However, there are two solutions to the problem of executing an incorrigible and dangerous deviant who cannot be coped with otherwise.

One is for the entire group (aside from his close kin) to attack and kill him collectively (Lee 1979), which ensures no one can be targeted for retaliation. The other is for a group consensus to be made known to a close kinsman such as a brother, who then executes him by ambush. In that case, the deviant's kin group has already suffered one loss and predictably it will not incur another by killing the executioner (Boehm 2011). In either event, the group rids itself of a serious social predator, and to the extent that this bully's potential victims were many, group members can share some serious reproductive gains by getting rid of such a predator. Thus, the costs of punishing capitally are underwritten by the gains of a rank and file who are no longer being exploited or despoiled.

Why Are There Any Bullies Left?

If serious bullies have been killed so regularly for thousands of generations, we must ask why bullying tendencies have remained so prominent in LPA hunter-gatherer social life, and why the gene pools they evolved have left us with such strong propensities in this direction. To answer this question, we must look to lesser forms of sanctioning that are not as ultimate as execution, and to special faculties humans have evolved in the area of personal self-control.

In assessing *nonultimate* sanctioning for these same 49 LPA foragers, I found there were ample opportunities for reform with respect to lesser deviances. A range of nonlethal sanctions are reported to stop deviant behavior, including strong central tendencies that favor methods such as gossip and public opinion as indirect sanctions, and direct sanctions such as ostracism and spatial distancing, verbal criticism, ridicule and shaming, and nonlethal physical punishment.

All these forms of social control incur reproductive losses, in that in the short run they tend to remove the deviant from relationships of cooperation. They also impair the reputation of the deviant, which can affect cooperative opportunities in the longer run, including marriage prospects. However, if a deviant reforms it is often possible to avoid further punishment and further reduction of reproductive success, and that person often may be able to restore a socially useful reputation (Boehm 2012b). In LPA foragers, this takes place in a context of symbolic and moral behavior, which means that groups have rules that are known by all with reference to bullying and theft—the latter being taken as a major type of cheating because deception is involved.

This evolved, rule-conscious, morally aware capacity response of many deviants to reform, just as their groups want them to, involves a special human capacity for self-awareness and self-control, which might be called an evolutionary conscience. As normally conceived, a conscience involves *rule internalization* (Gintis 2003) that is akin to what Simon (1990) was discussing under the rubric of cultural docility: one absorbs the rules of one's culture, although at the same time one is prone to break them.

However, a conscience enables a human being not only to identify with rules as a generalized curb on predatory behaviors, but also to respond directly to active disapproval because individuals are jealous of their future reputations, and also are fearful of serious or ultimate punishment. Although traditional views of the conscience have looked to it as an instrument of virtue, I would also incorporate Alexander's (1987) less rosy view that a conscience also allows us to reckon what we can get away with without adversely damaging our reproductive success.

At a gross level, similar inhibitions are involved when a great ape such as a chimpanzee or bonobo reckons the power of a competing individual or coalition and fearfully avoids competition that will have a negative payoff, but I emphasize that in these apes *fear* is the dominant element. Because of rule internalization, the human conscience permits far more subtle assessments that go well beyond fear. These include feelings of right and wrong and also concerns for the social reputation of the actor, and this provides a unique instrument for social navigation in a society that is not only judgmental and dangerously punitive, but also prone to reward virtuous individuals including altruists (Boehm 2012b).

The Egalitarian Syndrome

Our human gene pool was evolved most recently for us by LPA foragers, and these people seem to be concerned with bullying behavior as their main moral concern; so far, this has been demonstrated through statistical summaries of lethal sanctioning patterns, but the same patterning applies to my data on less-than-lethal sanctioning. It is also of interest that in a decades-long study of !Kung Bushmen, Wiessner (2005a, 2005b) found that when people originally living in nomadic bands talked about the social problems they faced, "big shot" behavior was the most prominent. If bullies are by far the main focus, the *cheaters* they punish seem to be more often thieves (Boehm 2012b) than the "tricksters" characterized by Williams (1966) and his many followers, starting with Trivers (1971).

Elsewhere (Boehm 1997), I have characterized LPA foragers in a way that explains their behavior patterns in terms of behavioral ecology, and a centerpiece for that discussion was their ecological niche as large-game hunters. Like other social carnivores such as lions, wolves, or dolphins, a major nutritional reliance on large carcasses brings with it the need to keep an entire hunting team nourished, and because such animals have social dominance hierarchies, there is the obvious problem that the alpha males can't monopolize the meat to a degree that lesser hunters are not properly nourished, which would make them unable to contribute effectively to team hunting efforts.

Good health and adequate energy are critical to a demanding subsistence task such as hunting (Stanford 1999), and this applies to the entire cooperative team. Thus, if a way is not found to share meat widely despite individuals having very different dominance positions, the entire team, including the alphas who are intent on monopolizing the meat, may be disadvantaged. In other social carnivores, the mechanism that spreads the meat around would appear to be a tolerance on the parts of dominants toward subordinates, when it comes to sharing, and although the sharing is not totally equitable, in general the subordinates do receive an adequate share and the group can stay in business (Boehm 2012b).

In humans, this tendency to share equitably is transformed by cultural factors such as symbolic communication, which makes it possible to define social behavior in moral terms. The result is a powerful egalitarian ethos which informs an entire egalitarian syndrome (Boehm 1997), with sharing of large game being a major concern, but personal freedom being equally salient. Richard Lee's (1979) often-cited and eloquent quotations from !Kung Bushmen exemplify these egalitarian attitudes, for so strong are group restrictions on self-aggrandizement that a highly successful hunter will diminish or even bad-mouth his own great accomplishments lest others think he is trying to use his hunting successes to lord it over others.

A more practical aspect of the egalitarian syndrome is that a successful hunter is not even allowed to preside over the carcass and its division; rather, in effect, large carcasses are regularly taken to be the "property" of the group (Boehm 2004), whereas neutral individuals are given the job of distributing the precious meat. This not only keeps the hunter from monopolizing meat for the consumption of his family or cronies but also prevents his using the meat for political purposes to increase his own clout in band affairs.

In some cases, the result is evenhanded distribution of meat to all the families in a band, but in others the successful hunter may be given a moderate meat bonus (Kelly 1995), perhaps as an inducement to continue his strenuous endeavors that, in the case of the more successful hunters, become highly altruistic. However, in all LPA groups the better hunter's compensatory payoffs also come in the form of reputational benefits (Kaplan and Hill 1985; Alexander 1987) as described earlier.

At the level of genes, the upshot is that the entire hunting team is being nourished adequately, as are the families of the hunters, whereas the best hunters' altruism is being compensated always through reputational benefits and sometimes through a larger share of meat. This syndrome applied strongly to culturally modern humans between 45,000 BP and 12,000 BP, but to the extent that equitable meat division is always a necessary concomitant of relying on large game, it may have applied to earlier humans, ever since they began to depend heavily on large ungulates for their subsistence. This began a quarter of a million years ago (Stiner 2002), with Archaic *Homo sapiens*.

The egalitarian syndrome has some significant precursors in Ancestral *Pan*. With chimpanzees and bonobos being the least common denominators, this ancestor had subordinate coalitions that reduced alpha power to some degree and set the stage for the development of extreme, definitive egalitarianism as experienced by LPA foragers. Based on what chimpanzees and bonobos do as a least common denominator, this ancestor also did some hunting of small game with limited sharing of meat, which involved dominant possessors sharing mainly with their kin and cronies. What humans did, much later, was to develop ideologies that guided their social control and enabled them to shape their own societies significantly. The result was a far more equalized sharing of a highly nutritious commodity.

In this manner, they developed an egalitarian syndrome that they themselves appreciated because this benefited both group subsistence cooperation and the personal autonomy of individuals, and because, more specifically, this enabled families to have a steady supply of meat. To do so they had to curb their meat-greedy alpha males who were born dominators, and sometimes they had to do so severely or even lethally.

Why Bullying Tendencies Continue to Be Robust

In egalitarian LPA foraging bands, active bullies were not tolerated, and if they could not control themselves they were punished severely. So why did tendencies to bully continue even after minimally 2000 generations of severe alpha suppression and, more likely, something approaching 10,000 generations? Even the minimum figure provided ample time for the underlying behavioral dispositions to be changed by natural selection, especially when the selection was of a social type that was highly focused by human intentions (Boehm 2008b).

My answer to this question is that the very efficiency of the human conscience may be responsible for this continuation. I believe that systems of social control provide ample opportunities for personal reform, and this means that a highly competitive alpha-type who also has a very efficient conscience can channel his (or her) aggressions in directions that will bring a net reproductive benefit, which means avoiding competitive acts that will bring on sanctioning. In this context, it is worth noting that despite their egalitarianism, the lifestyle of LPA foragers includes a substantial amount of individual competition, for only certain kinds of individual competition are curbed by the group; individuals may compete freely for spouses and for subsistence partners, and for many kinds of prestige (Fried 1967).

It is arrogant self-aggrandizement that is strongly disapproved and punished severely, precisely because people know where this can lead. And it is individuals whose personal self-control in this area is inadequate that pay the heavy reproductive penalties. We are speaking of functions of the conscience, here, but keep in mind that the conscience is evolutionary as well as moral, and one of its important functions is to enable expedient personal reform by socially sensitive individuals who are able to curb their impulses. At first blush, it may seem counterintuitive, but this capacity for reform could have enabled dominance tendencies to be preserved at fairly high levels in human gene pools—even after the development of strongly moralistic egalitarian behavior that led to harsh and systematic punishment of certain types of competitive dominance.

This continuation of dominance tendencies is quite evident in what has taken place since humans in the Holocene Epoch began to live in population densities that called for strong command and control. Knauft (1991) has defined the enigma of species that had long lived as equals suddenly turning hierarchical in the Holocene, and the first major surprise was the rapid rise of chiefdoms with their pronounced social classes and unabashedly authoritative leaders; next came early civilizations that were invariably despotic. Add this up, and there has to have been plenty of dominance potential left in our behavioral repertoire.

What I suggest, here, is that our many generations of effectively suppressing bullying free-riders at the level of phenotype would have reduced and probably transformed our political potential for dominance behavior in comparison with our ancestral starting point, which was more in line with what chimpanzees and bonobos do (Wrangham and Peterson 1996). Obviously, individuals with too-strong despotic tendencies and/or those with too-weak consciences have been culled as a result of egalitarian punishment, and this accounts for people having evolved consciences and for changes in our political dispositions.

Exactly how our political behavior has been transformed remains, however, a question for future discussion—although Wrangham and Peterson (1996) certainly have taken the subject of males and their demonic dispositions a long way. The fact that bullying behavior among children today has become such a topic of concern is another testimonial to the fact that executing bullies for thousands of generation obviously has failed to *eradicate* such behavior tendencies, and probably has transformed them.

Implications for Social Selection and Group Selection

As discussed earlier, genetic altruism has a number of possible mechanical explanations. It is possible that all of them have been involved in making cooperative humans relatively selfless and empathetic, as opposed to being extremely selfishly competitive. However, I would propose two of the seven types of selection mechanisms as mainstays in keeping our competitive species as altruistic as it is.

Perhaps the lesser is *group selection*. Human foragers have cooperated with their own group interests explicitly in mind (Boehm 2008b), and at the same time they have striven to curb selfishly predatory behavior that damages group functioning—even as altruistic cooperativeness has been actively promoted. In this context, foraging bands are sized appropriately for group selection to operate, and although such groups are quite porous in general (Palmer et al. 1997), intergroup conflict could have given an impetus to group selection in the Late Pleistocene (Bowles 2006).

Social selection is a newer and less explored entry, which takes place *within* groups. Such selection is based on both negative and positive social preferences because they affect partner choice by individuals and because they influence or determine the targets for social sanctioning. Such punishment has strong effects on reproductive success not only when capital punishment is used but also when access to cooperative relationships is impaired by lesser sanctions. Group preferences channel both of these social modes of selection, and much of the evolution of altruism in our species may be the result of them as they act in combination.

There is a significant connection between these two modes of selection, for a major detriment to *group selection* is the free-rider, especially the powerful bully but also the deceptive thief, who defects when others are staying a cooperative course. *Social selection* can support group selection not only by promoting and rewarding the individual altruism that abets cooperation but also by curbing the social predators who would otherwise take free rides. I propose that these two types of selection, working in tandem, may go a long way in explaining the genetic paradox of human altruism.

References

Alexander, Richard D. 1987. *The biology of moral systems*. New York: Aldine de Gruyter.

Betzig, Laura L. 1986. *Despotism and differential reproduction: A Darwinian view of history*. New York: Aldine.

Binford, Lewis. 2001. *Constructing frames of reference: An analytical method for archeological theory building using hunter–gatherer and environmental data sets*. Berkeley: University of California Press.

Boehm, Christopher. 1978. Rational pre-selection from *Hamadryas* to *Homo sapiens*: The place of decisions in adaptive process. *American Anthropologist* 80: 265–296.

———. 1981. Parasitic selection and group selection: A study of conflict interference in rhesus and Japanese macaque monkeys. In A. B. Chiarelli and R. S. Corruccini, eds., *Primate behavior and sociobiology: Proceedings of the International Congress of Primatology*, 161–182. Heidelberg: Springer.

———. 1984. Can hierarchy and egalitarianism both be ascribed to the same causal forces? *Politics and the Life Sciences* 1: 34–37.

———. 1993. Egalitarian behavior and reverse dominance hierarchy. *Current Anthropology* 34: 227–254.

———. 1997. Impact of the human egalitarian syndrome on Darwinian selection mechanics. *American Naturalist* 150: 100–121.

———. 1999. *Hierarchy in the forest: The evolution of egalitarian behavior*. Cambridge, MA: Harvard University Press.

———. 2000. Conflict and the evolution of social control. *Journal of Consciousness Studies* 7: 79–183. (special issue on evolutionary origins of morality [L. Katz, guest editor]).

———. 2004. What makes humans economically distinctive? A three-species evolutionary comparison and historical analysis. *Journal of Bioeconomics* 6: 109–135.

———. 2008a. A biocultural evolutionary exploration of supernatural sanctioning. In Joseph Bulbulia, Richard Sosis, Russell Genet, Ereika Harris, Karen Wyman, and Cheryl Genet, eds., *Evolution of religion: Studies, theories and critiques*, 143–152. Santa Margarita: Collins Foundation Press.

———. 2008b. Purposive social selection and the evolution of human altruism. *Cross-Cultural Research* 42: 319–352.

———. 2011. A natural history of retaliatory violence. *British Journal of Criminology* 51: 518–534.

———. 2012a. Ancestral hierarchy and conflict. *Science* 336: 844–847.

———. 2012b. *Moral origins: The evolution of virtue, altruism, and shame*. New York: Basic Books.

Bowles, Samuel. 2006. Group competition, reproductive leveling, and the evolution of human altruism. *Science* 314: 1569–1572.

Campbell, Donald T. 1975. On the conflicts between biological and social evolution and between psychology and moral tradition. *American Psychologist* 30: 1103–1126.

Cosmides, Leda, John Tooby, Laurence Fiddick, and Gregory A. Bryant. 2005. Detecting cheaters. *Trends in Cognitive Sciences* 9: 505–506.

Darwin, Charles. 1871. *The descent of man and selection in relation to sex*. London: John Murray.

de Waal, Frans B. M. 1982. *Chimpanzee politics: Power and sex among apes*. New York: Harper and Row.

———. 1996. *Good natured: The origins of right and wrong in humans and other animals*. Cambridge. MA: Harvard University Press.

de Waal, Frans B. M., and Frans Lanting. 1997. *Bonobo: The forgotten ape*. Berkeley: University of California Press.

Ellis, Lee. 1995. Dominance and reproductive success among nonhuman animals: A cross-species comparison. *Ethology and Sociobiology* 16: 257–333.

Fried, Morton H. 1967. *The evolution of political society*. New York: Random House.

Gintis, Herbert. 2003. The hitchhiker's guide to altruism: Gene-culture coevolution and the internalization of norms. *Journal of Theoretical Biology* 220: 407–418.

Goodall, Jane. 1992. Unusual violence in the overthrow of an alpha male chimpanzee at Gombe. In Toshishada Nishida, William C. Mcgrew, Peter Marler, Martin Pickford, and Frans B. M. De Waal, eds., *Topics in primatology, Volume 1: Human origins*, 131–142. Tokyo: University of Tokyo Press.

Hamilton, William D. 1964. The genetical evolution of social behavior I, II. *Journal of Theoretical Biology* 7: 1–52.

Harcourt, Alexander H., and Frans B. M. de Waal, eds. 1992. *Us against them: Coalitions and alliances in humans and other animals*. New York: Oxford University Press.

Hardin, Garrett. 1968. The tragedy of the commons. *Science* 162: 1243–1248.

Hill, Kim R., Robert Walker, Miran Božičević, James Eder, Thomas Headland, Barry Hewlett, A. Magdalena Hurtado, Frank Marlowe, Polly Wiessner, and Brian Wood. 2011. Coresidence patterns in hunter-gatherer societies show unique human social structure. *Science* 331: 1286–1289.

Hohmann, Gottfried, and Barbara Fruth. 2011. Is blood ticker than water? In Martha M. Robbins and Christoph Boesch, eds., *Among African apes*, 61–76. Berkeley: University of California Press.

Kaplan, Hillard, and Kim Hill. 1985. Hunting ability and reproductive success among male Aché foragers: Preliminary results. *Current Anthropology* 26: 131–133.

Kelly, Robert L. 1995. *The foraging spectrum: Diversity in hunter-gatherer lifeways*. Washington, DC: Smithsonian.

Klein, Richard G. 1999. *The human career: Human biological and cultural origins*. Chicago: University of Chicago Press.

Knauft, Bruce M. 1991. Violence and sociality in human evolution. *Current Anthropology* 32: 391–428.

Lee, Richard B. 1979. *The !Kung San: Men, women, and work in a foraging society*. Cambridge: Cambridge University Press.

Lion, Sebastien., Vincent A. A. Jansen, and Troy Day. 2011. Evolution in structured populations: Beyond the kin versus group debate. *Trends in Ecology & Evolution* 26: 193–201.

Lorenz, Konrad. 1966. *On aggression*. New York: Bantam.

McBrearty, Sally, and Alison Brooks. 2000. The revolution that wasn't: A new interpretation of the origin of modern human behavior. *Journal of Human Evolution* 39: 453–563.

Mitani, John C. 2005. Reciprocal exchange in chimpanzees and other primates. In P. Kappeler and Carel Van Schaik, eds., *Cooperation in primates: Mechanisms and evolution*, 101–113. Heidelberg: Springer-Verlag.

Nesse, Randolph M. 2007. Runaway social selection for displays of partner value and altruism. *Biological Theory* 2: 143–155.

Nishida, Toshisada. 1996. The death of Ntologi, the unparalleled leader of M Group. *Pan Africa News* 3: 4.

Palmer, Craig T., B. Eric Fredrickson, and Christopher F. Tilley. 1997. Categories and gatherings: Group selection and the mythology of cultural anthropology. *Evolution and Human Behavior* 18: 291–308.

Parker, Ian. 2007. Swingers: Bonobos are celebrated as peace-loving, matriarchal, and sexually liberated. Are they? *New Yorker*, July, 48–61.

Pilot, Malgorzata. 2005. Altruism as an advertisement: A model of the evolution of cooperation based on Zahavi's handicap principle. *Ethology, Ecology and Evolution* 17: 217–231.

Ruvolo, Maryellen, Todd R. Disotell, Mark W. Allard, Wesley M. Brown, and Rodney L. Honeycutt. 1991. Resolution of the African hominoid trichotomy by use of a mitochondrial gene sequence. *Proceedings of the National Academy of Science United States of America* 88: 1570–1574.

Schjelderup-Ebbe, Thorleif. 1935. Social behavior of birds. In Carl Murchison, ed., *A handbook of social psychology*, 947–972. Worcester, MA: Clark University Press.

Simon, Herbert A. 1990. A mechanism for social selection and successful altruism. *Science* 250: 1665–1668.

Stanford, Craig B. 1999. *The hunting apes: Meat eating and the origins of human behavior*. Princeton: Princeton University Press.

Stiner, Mary C. 2002. Carnivory, coevolution, and the geographic spread of the genus *Homo*. *Journal of Archaeological Research* 10: 1–63.

Trivers, Robert L. 1971. The evolution of reciprocal altruism. *Quarterly Review of Biology* 46: 35–57.

Vehrencamp, Sandra L. 1983. A model for the evolution of despotic versus egalitarian societies. *Animal Behavior* 31: 667–682.

Watts, David P., and John C. Mitani. 2002. Hunting and meat sharing by chimpanzees at Ngogo, Kibale National Park, Uganda. In Cristoph Boesch, Gottfried Hohmann, and Linda Marchant, eds., *Behavioral diversity in chimpanzees and bonobos*, 244–255. Cambridge: Cambridge University Press.

West, Stuart. A, Ashleigh S. Griffin, and Andy Gardner. 2007. Social semantics: Altruism, cooperation, mutualism, strong reciprocity and group selection. *Journal of Evolutionary Biology* 20: 415–432.

West-Eberhard, Mary Jane. 1975. The evolution of social behavior by kin selection. *The Quarterly Review of Biology* 50: 1–33.

———. 1979. Sexual selection, social competition, and evolution. *Proceedings of the American Philosophical Society* 123: 222–234.

Wiessner, Polly. 2005a. Norm enforcement among the Ju/'hoansi Bushmen: A case of strong reciprocity? *Human Nature* 16: 115–145.

———. 2005b. *Verbal criticism: Ju/'hoansi style punishment. http://www.peacefulsocieties.org/NAR06/060105juho.html* [Online].

Williams, George C. 1966. *Adaptation and natural selection: A critique of some current evolutionary thought.* Princeton: Princeton University Press.

Wilson, David S. 1975a. A theory of group selection. *Proceedings of the National Academy of Sciences of the United States of America* 72: 143–146.

———. 1999. A critique of R. D. Alexander's views on group selection. *Biology and Philosophy* 14: 431–449.

Wilson, David S., and Edward O. Wilson. 2007. Rethinking the theoretical foundation of sociobiology. *Quarterly Review of Biology* 82: 327–348.

Wilson, Edward O. 1975b. *Sociobiology: The new synthesis.* Cambridge, MA: Harvard University Press.

Wrangham, Richard W. 1987. African apes: The significance of African apes for reconstructing social evolution. In Warren G. Kinzey, ed., *The evolution of human behavior: Primate models,* 51–71. Albany: SUNY Press.

Wrangham, Richard, and Dale Peterson. 1996. *Demonic males: Apes and the origins of human violence.* New York: Houghton Mifflin.

Wynne-Edwards, Vero C. 1962. *Animal dispersion in relation to social behavior.* Edinburgh: Oliver and Boyd.

Zahavi, Amotz. 1995. Altruism as a handicap: The limitations of kin selection and reciprocity. *Journal of Avian Biology* 26: 1–3.

Zahavi, Amotz, Avishag Zahavi, Naama Zahavi-Ely, and Melvin Patrick Ely. 1997. *The handicap principle: A missing piece of Darwin's puzzle.* New York: Oxford University Press.

3 THE STRUCTURE AND EVOLUTION OF MORALITY

PUBLIC AND PRIVATE PERSONA

Herbert Gintis

Behavioral morality is the set of moral rules we attribute to people by virtue of their actions. *Classical morality* is the set of rules that philosophers and theologians consider that moral individuals are obliged to obey. The content of both behavioral and classical morality are contested, and the appropriate relationship between the two is relatively complex. This chapter deals with behavioral morality alone.

Traditional social science embraces a rather straightforward understanding of the relationship between human biological evolution and morality. This is the venerable notion of *tabula rasa* so ably critiqued by evolutionary psychologists (Tooby and Cosmides 1992; Pinker 2002). According to this story, evolution gave humans large brains. The brain is empty at birth but filled with culture, including moral principles, by society. Beyond providing us with the cognitive capacity to understand moral principles, we are told, biological evolution has nothing to do with morality.

According to this story, behavioral morality is the pure product of *cultural evolution*. This idea is famously expressed by Thomas Hobbes (1968 [1651]), who wrote:

> The state of men without civil society (which may be called the state of nature) is nothing but a war of all against all. . . . Where every man is enemy to every man, the life of man is solitary, poor, nasty, brutish, and short.

Some two centuries later, the influential economist Francis Ysidro Edgeworth (1881) affirmed:

> The first principle of economics is that every agent is actuated only by self-interest.

We find the same sentiment a century later in prominent biologist Richard Dawkins (1976), who wrote:

> We are survival machines—robot vehicles blindly programmed to preserve the selfish molecules known as genes. Let us try to *teach* generosity and altruism, because we are born selfish.

The general conclusion, then, is that humans are inherently egoistic, but we can *teach* the young the proper way to behave and we can *enforce* proper behavior by means of civilizing institutions. Morality, then, is an elaborate veneer hiding our basic self-regarding natures. The unusually provocative Michael Ghiselin (1974) wrote:

> No hint of genuine charity ameliorates our vision of society, once sentimentalism has been laid aside. What passes for cooperation turns out to be a mixture of opportunism and exploitation. Scratch an altruist, and watch a hypocrite bleed.

More dispassionately, noted evolutionary biologist Richard Alexander (1987) professed:

> Ethics, morality, human conduct, and the human psyche are to be understood only if societies are seen as collections of individuals seeking their own self-interest.

This chapter presents a rather more sanguine appreciation of behavioral morality, an appreciation based on evolutionary biology, paleontology, the rational actor model, behavioral game theory, and experimental psychology. The basic principles are as follows:

- Behavioral morality is the product of an evolutionary dynamic extending over hundreds of thousands of years in the hominin line involving the interaction of genes and culture.
- In this dynamic, hominin societies transformed culture, and the new culture made new behaviors fitness enhancing, transforming the hominin line itself.

Thus, *gene-culture coevolution*: in humans, genes are the product of culture and culture is the product of genes.

- Behavioral morality, in particular, is predicated on a set of human *predispositions* that evolved during our evolutionary emergence in small-scale hunter-gatherer groups.
- When our ancestors developed the capacity intentionally to devise social games and play according to their culturally constituted rules, it became possible to conceive of society itself as a social game, the rules of which are determined in a new arena of social life, which we may call the *public sphere*.
- Humans thus evolved two modes of social behavior: *private persona*, personal preferences regulating everyday life in civil society; and their *public persona*, regulating their behavior in the public sphere.
- At the heart of our moral capacities, both as private persona and public persona, is the capacity to conceptualize a *higher moral realm* that leads us to "do the right thing," to feel the satisfaction of doing the right thing, and to experience a degraded self when we have not done the right thing.

Gene-Culture Coevolution

Individual fitness in humans depends on the structure of social life. For instance, if society rewards certain behaviors, then females will prefer offspring who exhibit these behaviors, and if there is a genetic element in the behaviors, they will seek mates who exhibit these behaviors as well. Thus, social values entail enhanced fitness for males that carry the socially valued genes. Similarly, if social norms entail the ostracism of individuals who exhibit certain behaviors, then genes that support these behaviors are likely to be replaced in the population by genes that suppress the disfavored behaviors.

Human cognitive, affective, and moral capacities are thus the product of an evolutionary dynamic involving the interaction of genes and culture. We call this dynamic *gene-culture coevolution* (Cavalli-Sforza and Feldman 1982; Boyd and Richerson 1985; Dunbar 1993; Richerson and Boyd 2004). This co-evolutionary process has endowed us with preferences that go beyond the self-regarding concerns emphasized in traditional economic and biological theory, with a social epistemology that facilitates the sharing of intentionality across minds, and a moral sense that entails both contributing to the social good and doing the right thing for its own sake. Gene-culture coevolution explains the salience of such other-regarding values as a taste for cooperation, fairness, and retribution; the capacity to empathize; and the ability to value such character virtues as honesty, hard work, piety, and loyalty.

Gene-culture coevolution is the application of *sociobiology*, the general theory of the social organization of biological species, to humans—a species that transmits culture in a manner that leads to its preservation across many generations.

The genome in general encodes information that is used both to construct a new organism and to endow it with instructions for transforming sensory inputs into behavioral outputs. Because learning is costly, efficient information transmission will ensure the genome encodes those aspects of the organism's environment that are *constant*, or that change only very slowly through time and space compared with the reproduction period for the species. In contrast, environmental conditions that vary rapidly can be dealt with by providing the organism with sufficient phenotypic plasticity.

There is an intermediate case, however, that is handled efficiently neither by genetic encoding nor phenotypic plasticity. When environmental conditions are correlated positively but imperfectly across generations, each generation acquires valuable information through *learning*. In such cases there is a fitness benefit to information transmission through nongenetic channels. Several such transmission mechanisms have been identified (Jablonka and Lamb 1995), among which cultural transmission (Bonner 1984; Richerson and Boyd 1998) is a most flexible form.

Cultural transmission takes the form of parents to children, peer to peer, and elder to younger (Cavalli-Sforza and Feldman 1981); higher to lower status (Henrich and Gil-White 2001); popularity (Newman et al. 2006), and even random (Shennan 1997; Skibo and Bentley 2003). The similarity of cultural and biological evolution goes back to Julian Huxley (1955), Karl Popper (1979), and William James (1880). The idea of treating culture as a form of epigenetic transmission was pioneered by Dawkins (1976), who coined the term *meme* in *The Selfish Gene* (1976). There quickly followed several major contributions to a biological approach to culture, all based on the notion that culture, like genes, could evolve through replication, mutation, and selection. Culture propagates from brain to brain, mutates in replication, and is subject to selection according to its effects on the fitness of its carriers (Parsons 1964; Cavalli-Sforza and Feldman 1982).

There are strong interactions between genes and culture, ranging from basic physiology, such as the transformation of the organs of speech with the evolution of language, to sophisticated social emotions, including empathy, shame, guilt, and revenge-seeking (Zajonc 1980; Ihara 2011). Because of their common informational and evolutionary character, there are strong parallels between models of genetic and cultural evolution (Mesoudi et al. 2006). Like genes, culture is transmitted from parents to offspring, and, like culture, which is transmitted horizontally to unrelated individuals, so are genes in microbes and many plant species; genes are regularly transferred across lineage boundaries (Jablonka and Lamb 1995; Abbott et al. 2003; Rivera and Lake 2004).

Anthropologists reconstruct the history of social groups by analyzing homologous and analogous cultural traits, much as biologists reconstruct the evolution of species by the analysis of shared characters and homologous DNA (Mace and Pagel 1994). Indeed, the same computer programs developed by biological systematists are used by cultural anthropologists (Holden 2002; Holden and Mace 2003). In addition, archeologists who study cultural evolution have a modus operandi similar to paleobiologists who study genetic evolution (Mesoudi et al. 2006). Both attempt to reconstruct lineages of artifacts and their carriers. Like paleobiologists, archaeologists assume that when analogy can be ruled out, similarity implies causal connection by inheritance (O'Brien and Lyman 2000). Like biogeographers' study of the spatial distribution of organisms (Brown and Lomolino 1998), behavioral ecologists study the interaction of ecological, historical, and geographic factors that determine distribution of cultural forms across space and time (Winterhalder and Smith 1992).

Perhaps the most common criticism of the analogy between genetic and cultural evolution is that the gene is a well-defined, discrete, independently reproducing, and mutating entity, whereas the boundaries of the unit of culture are ill-defined and overlapping. However, this view of the gene is outdated. We now know that overlapping, nested and movable genes have some of the fluidity of cultural units, whereas quite often the boundaries of a cultural unit (a belief, icon, word, technique, stylistic convention) are quite delimited and specific. Similarly, alternative splicing, nuclear and messenger RNA editing, cellular protein modification, and genomic imprinting, which are quite common, undermine the standard view of the insular gene coding for a single protein and support the notion of genes having variable boundaries and strongly context-dependent effects. Moreover, natural selection requires heritable variation and selection, but does not require discretely transmitted units.

Dawkins (1982) added a second fundamental mechanism of epigenetic information transmission, noting that organisms can transmit environmental artifacts directly to the next generation in the form of such constructs as beaver dams, beehives, and even social practices (such as mating rituals and hunting strategies) (Odling-Smee et al. 2003; Gintis 2014).

An excellent example of gene-environment coevolution is the honeybee, in which the origin of its eusociality probably lay in a high degree of relatedness, but which persists in modern species despite the fact that relatedness in the hive is generally quite low as a result of multiple queen matings, multiple queens, queen deaths, and the like (Gadagkar 1991; Seeley 1997; Wilson and Hölldobler 2005).

The social structure of the hive, a classic example of niche construction, is transmitted genetically across generations, and the honeybee genome is an adaptation to the social structure of the hive laid down in the distant past.

Gene-culture coevolution in humans is a special case of gene-environment coevolution in which the environment is culturally constituted and transmitted (Feldman and Zhivotovsky 1992). The key to the success of our species in the framework of the hunter-gatherer social structure in which we evolved is the capacity of genealogically unrelated individuals to cooperate in large egalitarian groups in hunting, territorial acquisition, and defense (Boehm 1999; Richerson and Boyd 2004). Although some contemporary biological and economic theorists have attempted to show that such cooperation can be supported by self-regarding rational agents (Trivers 1971; Alexander 1987; Fudenberg et al. 1994), the conditions under which their models work are implausible even for small groups (Boyd and Richerson 1988; Gintis 2009; Bowles and Gintis 2011). Rather, the social environment of early humans was conducive to the development of prosocial traits, such as empathy, shame, pride, embarrassment; and reciprocity, without which social cooperation would be impossible (Sterelny 2011).

Neuroscientific studies exhibit clearly the genetic basis for moral behavior. Brain regions involved in moral judgments and behavior include the prefrontal cortex, the orbitofrontal cortex, and the superior temporal sulcus (Moll et al. 2005). These brain structures are virtually unique to or most highly developed in humans and are doubtless evolutionary adaptations (Schulkin 2000). The evolution of the human prefrontal cortex is tied closely to the emergence of human morality (Allman et al. 2002). Patients with focal damage to one or more of these areas exhibit a variety of antisocial behaviors, including the absence of embarrassment, pride, and regret (Beer et al. 2003; Camille 2004), as well as sociopathic behavior (Miller et al. 1997). There is a probable genetic predisposition underlying sociopathy, and sociopaths comprise 3% to 4% of the male population, but they account for between 33% and 80% of the population of chronic criminal offenders in the United States (Mednick et al. 1977). It is clear from this body of empirical information that culture is encoded directly into the human brain with symbolic representations in the form of cultural artifacts. This, of course, is the central claim of gene-culture co-evolutionary theory.

Culture to Genes: The Physiology of Communication

The evolution of the physiology of speech and facial communication is a dramatic example of gene-culture coevolution. The increased social importance of communication in human society rewarded genetic changes that facilitate speech. Regions in the motor cortex expanded in early humans to facilitate speech production. Concurrently, nerves and muscles to the mouth, larynx, and tongue became more numerous to handle the complexities of speech (Jurmain et al. 1997). Parts of the cerebral cortex, Broca's and Wernicke's areas, which do not exist or are relatively small in other primates, are large in humans and permit grammatical speech and comprehension (Binder et al. 1997; Belin et al. 2000).

Modern humans have a larynx low in the throat, a position that allows the throat to serve as a resonating chamber capable of a great number of sounds (Relethford 2007). The first hominins that have skeletal structures supporting this laryngeal placement are the *Homo heidelbergensis*, who lived from 800,000 years to 100,000 years ago. In addition, the production of consonants requires a short oral cavity. Our nearest primate relatives have much too long an oral cavity for this purpose. The position of the hyoid bone, which is a point of attachment for a tongue muscle, developed in *Homo sapiens* in a manner permitting highly precise and flexible tongue movements.

Another indication that the tongue has evolved in hominins to facilitate speech is the size of the hypoglossal canal, an aperture that permits the hypoglossal nerve to reach the tongue muscles. This aperture is much larger in Neanderthals and humans than in early hominins and nonhuman primates (Dunbar 2005). Human facial nerves and musculature have also evolved to facilitate communication. This musculature is present in all vertebrates, but in mammals it serves feeding and respiratory functions alone (Burrows 2008). In mammals, this mimetic musculature attaches to the skin of the face, thus permitting the facial communication of such emotions as fear, surprise, disgust, and anger. In most mammals, however, a few wide, sheetlike muscles are involved, rendering fine information differentiation impossible, whereas in primates this musculature divides into many independent muscles with distinct points of attachment to the epidermis, thus permitting higher bandwidth facial communication. Humans have the most highly developed facial musculature by far of any primate species, with a degree of involvement of lips and eyes that is not present in any other species.

In short, humans have evolved a highly specialized and very costly array of physiological characteristics that both presuppose and facilitate sophisticated vocal and visual communication, whereas communication in other primates, lacking as they are in cumulative culture, goes little beyond simple calling and gesturing capacities involving adoption of communicative physiology. This example is quite a dramatic and concrete illustration of the intimate interaction of genes and culture in the evolution of our species.

The Rationality of Morality

Behavioral morality involves making personally costly choices that promote ethical goals. People not only balance self-regarding against moral concerns, but also face conflicting moral principles in making choices. We therefore model choice behavior using the *rational actor model*, according to which individuals have a *preference function* representing their goals, they face *constraints* that limit the mix of goals available to them, and they have *beliefs* concerning how their actions affect the probability of attaining their goals. This concept of rationality is extremely skeletal, strongly favoring the consistency principles of *formal rationality* with little regard for the actors' *substantive rationality*—in other words, the extent to which behavior is attuned to achieving any particular standard measure, such as fitness, well-being, or happiness. Preferences, for instance, may include such *self-regarding* goals as material wealth and leisure, such *other-regarding* goals as fairness and consideration for the welfare of others, and such *character virtues* as honesty, loyalty, trustworthiness, courage, and considerateness that have intrinsic value independent of their effects. Moreover, we impose no plausibility constraints on beliefs.

The rational actor model *assumes* but does not *explain* the pattern of individual preferences. Understanding preferences requires forays into the psychology of goal-directed, moral behavior (Haidt 2012), social evolutionary theory (Tooby and Cosmides 1992), and problem-solving heuristics (Gigerenzer and Todd 1999).

The most important single contribution to the theory of formal rational choice was that of Leonard Savage (1954), who showed that a small set of plausible choice axioms (the *Savage axioms*) implies that a rational actor can be modeled as maximizing an objective function subject to the constraints he faces, where his beliefs take the form of a *subjective prior* specifying the agent's judgment regarding the probabilistic effects of his actions on the attainment of his goals. This objective function is often called a *utility function*, although the term is misleading because the objective function in the rational actor model need have no utilitarian content. The most important of the Savage axioms is that the agent's preferences are *transitive* in the sense that if he prefers A to B and he also prefers B to C, then he must also prefer A to C. The remaining assumptions are rather technical and not relevant for our purposes (Savage 1954; Gintis 2009; Gintis and Helbing 2015).

The Savage axioms do not suggest that an agent chooses what is in his best interest or what gives him pleasure. Nor do the axioms suggest that the actor is selfish, calculating, or amoral. Finally, the Savage axioms do not suggest that the rational actor is trying to maximize utility or anything else. The maximization formulation of rational choice behavior is simply an analytic convenience, akin to the least action principle in classical mechanics, or predicting the behavior of an expert billiards player by solving a set of differential equations of which the expert has not the least awareness. The theory flowing from the Savage axioms is a powerful tool that is valid whatever the nature of human goals and motivations, provided they involve consistent choices.

A Typology of Rational Action

Human actors exhibit three types of motives in their daily lives: *self-regarding*, *other-regarding*, and *universalist*. Self-regarding motives include seeking wealth, consumption, leisure, social reputation, status, esteem, and other markers of personal advantage. Other-regarding motives include reciprocity, fairness, and concern for furthering the well-being of others. Universalist motives are those that are followed for their own sake rather than directly for their effects. Among these universalist goals, which we term *character virtues*, are honesty, loyalty, courage, trustworthiness, and considerateness. Of course, such universalist goals normally have consequences for those with whom one interacts, and for society as a whole. But one undertakes universalist actions *for their own sake*, beyond any consideration of their effects. I will give one example of other-regarding behavior and another of universalist behavior, as revealed by laboratory experiments using behavioral game theory.

Positive Reciprocity: The Trust Game

Positive reciprocity takes the form of individual responding to an act of kindness by returning the kindness. Positive reciprocity can be self-regarding because returning favors helps create and sustain a mutually rewarding relationship. Robert Trivers (1971) called such tit-for-tat behavior *reciprocal altruism*, but there is in fact no altruism at all involved, because a purely selfish individual will engage in this form of positive reciprocity. However, humans also exhibit positive reciprocity when there is no possibility of future gain from the costly act of returning a kindness. We call this other-regarding behavior *positive reciprocity*, or *altruistic cooperation*.

For example, consider the *trust game*, first studied by Berg et al. (1995). In this game, carried out in an experimental laboratory, subjects are each given an endowment, say $10. Subjects are then randomly paired, and one subject in each pair, whom we will call the Proposer, is told he can transfer any number of dollars, from $0 to $10, to his anonymous partner, whom we will call the Respondent, and the Proposer can keep the remainder. The amount transferred will be *tripled* by the experimenter and given to the Respondent, who can then give any number of dollars back to the Proposer (this amount is not tripled). A Proposer who transfers a lot is called *trusting*, and a Respondent who returns a lot to the Proposer is called *trustworthy*. This interaction occurs only one time, and the Proposer and the Respondent never learn each other's identity. Trustworthiness is thus a pure act of other-regarding positive reciprocity.

Berg et al. (1995) found that, on average, the Proposer transferred $5.16 of the $10.00 to the Respondent, and on average, the Respondent transferred back $4.66 the Proposer. Furthermore, when the experimenters revealed this result to the subjects and had them play the game a second time, on average $5.36 was transferred from the Proposer to the Respondent, and $6.46 was transferred back from the Respondent to the Proposer. In both sets of games there was a great deal of variability, some the Proposers transferring everything, some nothing, and some Respondents more than fully repaying their Proposers and others returning nothing.

Negative Reciprocity: The Ultimatum Game

Negative reciprocity occurs when an individual responds to an unkind act by retaliating with another unkind act. Negative reciprocity can be self-regarding because retaliation may induce the other person to behave more kindly in the future, and more generally one may thereby enhance one's reputation as someone not to be trifled with. There is no moral element in this sort of negative reciprocity, because a purely selfish individual may retaliate to enhance his reputation and thereby deter future unkind acts. However, humans also exhibit negative reciprocity when there is no possibility of future interaction with the offender. We call this other-regarding negative reciprocity *altruistic punishment*.

The simplest game exhibiting altruistic punishment is the Ultimatum Game (Güth et al. 1982). Under conditions of anonymity, two subjects, whom we will call Alice and Bob, are shown a sum of money, say $10. Alice, called the Proposer, is instructed to offer any number of dollars, from $1 to $10, to Bob, who is called the Responder. Alice can make only one offer and Bob can either accept or reject this offer. If Bob accepts the offer, the money is split according to Alice's offer. If Bob rejects the offer, both players receive nothing. Alice and Bob, who are unknown to each other, do not interact again.

If Bob is self-regarding, he will accept anything he is offered. If Alice believes Bob is self-regarding, she will offer him the minimum amount ($1) and Bob will accept. However, when actually played, this self-regarding outcome is almost never observed or even approximated. In fact, under varying conditions and with varying amounts of money, Proposers routinely offer Responders very substantial amounts (50% of the total generally being the modal offer) and Responders frequently reject offers less than 30% (Güth and Tietz 1990; Camerer and Thaler 1995). Are these results culturally dependent? Do they have a strong genetic component or do all successful cultures transmit similar values of reciprocity to individuals? Roth et al. (1991) conducted the Ultimatum Game in four different countries (United States, the former Yugoslavia, Japan, and Israel) and found that, although the level of offers differed a small but significant amount in different countries, the probability of an offer being rejected did not. This indicates that both Proposers and Responders share the same notion of what is considered fair in that society, and that Proposers adjust their offers to reflect this common notion. When a much greater degree of cultural diversity is studied, however, large differences in behavior are found, reflecting different standards of what it means to be fair in different types of societies (Henrich et al. 2004).

Behavior in the Ultimatum Game conforms to the altruistic punishment model. Responders reject offers less than 40% to hurt an unfair Proposer. Proposers offer 50% because they are altruistic cooperators, or 40% because they fear rejection. To support this interpretation, we note that if the offers in an Ultimatum Game are generated by a computer rather than by the Proposer, and if Responders know this, low offers are rarely rejected (Blount 1995). This suggests that players are motivated by reciprocity, reacting to a violation of behavioral norms (Greenberg and Frisch 1972). Moreover, in a variant of the game in which a Responder rejection leads to the Responder getting nothing but allows the Proposer to keep the share he suggested for himself, Responders never reject offers, and Proposers make considerably smaller (but still positive) offers (Bolton and Zwick 1995). As a final indication that altruistic punishment motives are operative in this game, after the game is over, when asked why they offered more than the lowest possible amount, Proposers commonly said they were afraid Responders would consider low offers unfair and reject them. When Responders rejected offers, they usually claimed they want to punish unfair behavior. In all these experiments a significant fraction of subjects (about a quarter, typically) conformed to purely self-regarding preferences.

A Universalist Character Virtue: Honesty

Certain moral behaviors are *universalist* in the sense that one performs them, at least in part, because it is virtuous to do so, apart from any effects they have on oneself, others, or society in general. For instance, one can be *honest* in dealing with another agent without caring at all about the effect on the other agent, or even caring about the impact of honest behavior on society at large. Similarly, one can be courageous in battle because it is the right thing to do, independent from the effect of one's actions on winning or losing the battle.

A particularly clear example of the value of honesty is reported by Gneezy (2005), who studied 450 undergraduate participants paired off to play three games of the following form, all payoffs to which are of the form a/b, where player 1 (Alice) receives a and player 2 (Bob) receives b. In all games, Alice was shown two pairs of payoffs, A:($x; y$) and B:($z; w$) where x, y, z, and w are amounts of money with $x < z$ and $y > w$, so in all cases, B is better for Bob and A is better for Alice. Alice could then say to Bob, who could not see the amounts of money, either "Option A will earn you more money than option B," or "Option B will earn you more money than option A." The first game was A:(5,6) versus B:(6,5), so Alice could gain one by lying and being believed, while imposing a cost of one on Bob. The second game was A:(5,15) versus B:(6,5) so Alice could gain 10 by lying and being believed, while still imposing a cost of one on Bob. The third game was A:(5,15) versus B:(15,5), so Alice could gain 10 by lying and being believed, while imposing a cost of 10 on Bob.

Before starting play, the experimenter asked each Alice whether she expected her advice to be followed, inducing honest responses by promising to reward her if her guesses were correct. He found that 82% of Alices expected their advice to be followed (the actual result was that 78% of Bobs followed their Alice's advice). It follows that if Alices were self-regarding, they would always lie and recommend B to their Bob.

The experimenters found that, in game two, where lying was very costly to Bob and the gain to lying for Alice was small, only 17% of subjects lied. In game one, where the cost of lying to Bob was only one but the gain to Alice was the same as in game two, 36% lied. In other words, subjects were loathe to lie, but considerably more so when it was costly to their partner. In game three, when the gain from lying was large for Alice, and equal to the loss to Bob, fully 52% lied. This shows that many subjects are willing to sacrifice material gain to avoid lying in a one-shot, anonymous interaction, with their willingness to lie increasing with an increased cost of truth-telling to themselves, and decreasing with an increase in their partner's cost of being deceived. Similar results were found by Boles et al. (2000) and Charness and Dufwenberg (2006). Gunnthorsdottir et al. (2002) and Burks et al. (2003) have shown that a social–psychological measure of "Machiavellianism" predicts which subjects are likely to be trustworthy and trusting.

The Public Sphere

The social life of most species, including mating practices, symbolic communication, and power relations, is inscribed in its *core genome* and expressed in stereotypical form by its members (Gintis 2014). *Homo sapiens* is unique in adapting its social life in fundamental and deep-rooted ways to environmental challenges and opportunities (Richerson and Boyd 2004). This flexibility is based on two aspects of our mental powers. The first is our ability to *devise new rules of the game in social life* and to *base our social interaction* on these new rules. This capacity, absent in other species, makes us *Homo ludens*: Man the game player. This capacity is possessed even by very young children who invent, understand, and play games for fun. In adult life, this same capacity is exercised when people come together to erect, protect, and transform the social rules that govern their daily life. Broadly speaking, we can define *the public sphere* as the arena in which society-wide rules of the game are considered, and *politics* as the cooperative, conflictual, and competitive behaviors through which rules are established and individuals are assigned to particular public positions.

Humans evolved in hunter-gather societies consisting of a dozen families or so (Kelly 1995), in which political life was an intimate part of daily life, involving the sorts of self-regarding, other-regarding, and universalistic motivations described earlier. In particular, political activity was strongly *consequentialist*: a single individual could expect to make a difference to the outcome of a deliberation, a conflict, or a collaboration, so that our political morality developed intimately entwined with material interests and everyday consequentialist moral sentiments (Boehm 1999; Gintis et al. 2015).

As we move from small-scale hunter-gatherer societies to modern mass societies with millions of members, the public sphere passes from being intimately embedded in daily life to being a largely detached institutional arena, governed by complex institutions controlled by a small set of individuals, and over which most members have, at best, formal influence through the ballot box and, at worst, no formal influence whatever. Political activity in modern societies is thus predominately *nonconsequentialist*, meaning individuals do not base their choices on the effect of their actions on political outcomes. Except for a small minority of individuals contesting for personal power, the political choices of a single citizen affect public sphere outcomes with a probability very close to zero—sufficiently close that these choices cannot not be attributed to consequentialist motives, whether self-regarding, other-regarding, or universalist.

In large elections, the rational consequentialist agent will not vote because the costs of voting are positive and significant, but the probability that one vote will alter the outcome of the election is vanishingly small, and adding a single vote to the total of a winning candidate enhances the winner's political efficacy, at best, an infinitesimal amount (Downs 1957; Riker and Ordeshook 1968). Thus, the personal consequentialist gain from voting is too small to motivate behavior even for a committed other-regarding or universalist altruist (Hamlin and Jennings 2011). For similar reasons, if one chooses to vote, there is no plausible reason to vote on the basis of the impact of the outcome of the election on one's personal material gains or on the basis of the gains to the demographic and social groups to which one belongs, or even on the basis of consequentialist universal values. One vote simply makes no difference. It follows also that the voter, if rational and consequentialist, and incapable of personally influencing the opinions of more than a few others, will not bother to form opinions on political issues, because these opinions cannot affect the outcome of elections. Yet people do vote, and many do expend time and energy in forming political opinions. Although voters do appear to behave strategically (Fedderson and Sandroni 2006), their behavior does not conform to the rational consequentialist model (Edlin et al. 2007).

It also follows that rational consequentialist individuals will not participate in the sort of collective actions that are responsible for the growth in the world of representative and democratic governance, the respect for civil liberties, the rights of minorities and gender equality in public life, and the like. In the rational consequentialist model, only small groups aspiring for social dominance will act politically. Yet, modern egalitarian political institutions are the result of such collective actions (Bowles and Gintis 1986; Giugni et al. 1998). This behavior cannot be explained by a rational consequentialist model.

Except for professional politicians and socially influential individuals, electoral politics is a vast morality play to which our consequentialist models of the rational actor are a poor fit. Mancur Olson argued as much in his classic, *The Logic of Collective Action* (1965), but behavioral scientists have yet to come fully to grips with its iron-clad logic (Downs 1957; Hamlin and Jennings 2011).

Defenders of the rational consequentialist model (there are very few) might respond that voters *believe* their votes make a difference, however poorly this belief might survive logical scrutiny. Indeed, when asked why they vote, voters' commonly respond they are trying to help get one or another party elected to office. When apprised of the illogical character of that response, given that a single vote in a large election cannot make a difference, they commonly reply that there are in fact close national elections, where the balance is tipped in one direction or another by only a few hundred votes. When reminded that one vote will not affect even such close elections, the common reply is, "Well, if everyone thought like that, then no one would vote and we could not have a democracy."

Politically active and informed citizens appear to operate on the principle that voting and participating in collective actions are *highly valued nonconsequentialist behaviors*. This idea is difficult for people to articulate because the consequentialist versus nonconsequentialist distinction is not part of either common parlance or the specialized lexicon of political theory. However, most voters agree with statements such as, "My single vote won't make a difference, but if all concerned citizens vote our common concerns, we can make a difference." Of course it does not logically follow that one should vote according to standard decision theory because if "my single vote won't make a difference," then I still have no consequentialist reason for voting.

However, humans appear to follow a nonconsequentialist logic that may be summarized as *rule-consequentialism*: in public life, choose a rule that like-minded people might plausibly choose and, if followed by all of us, it will lead to the most desirable outcome (Harsanyi 1977; Hooker 2015; Roemer 2010). Rule-consequentialism explains why assertions such as "I am helping my candidate win by voting" and "I am helping promote democracy by demonstrating against the dictator" are literally correct. Because rule-consequentialism is so ingrained in our public persona, people untrained in traditional rational decision theory simply cannot understand the argument that it is irrational to vote or to participate in collective actions, even when they can be persuaded that their actions are nonconsequential.

Rule-consequentialism can also explain many stylized facts of voter behavior. First, when the cost of voting increases, fewer people vote. The rule here is something like, "My unusual personal situation means voting would be very costly to me today. I would not expect anyone in my position to vote, so I am comfortable with not voting." Second, it explains why voter turnout is higher when the issues to be decided have greater social impact. Third, it explains why turnout is higher when the election is expected to be close. Finally, it explains why, in a two-party election, turnout is likely to be higher among voters for the side that is not expected to win. Indeed, it is reasonable to speculate that rule-consequentialism leads voters to act in very large elections in much the same way they would in very small elections, although in very small elections consequentialist issues (such as self-interest) may trump the nonconsequentialist rule.

We conclude that the individual immersed in consequentialist everyday life expresses his *private persona*, whereas his behavior in the public sphere reveals his *public persona*. Individuals acting in the public sphere are, then, a different sort of animal, one which Aristotle (2002 [350 BC]) called *zoon politikon* in his *Nicomachean Ethics*.

Private and Public Persona

The concept of a nonconsequentialist public persona suggests a two-by-three categorization of human motivations, as presented in Table 3.1. In this table, the three columns represent three modes of social interaction. The *personal* mode is purely self-regarding whereas the *social* mode represents the agent as embedded in a network of significant social relations, and the *universal* represents the individual's realm of recognized suprasituational moral obligations. The two rows represent the agent's private persona of consequentialist social relations in civil society, and the agent's public persona of nonconsequentialist political relationships in the public sphere.

Homo economicus is the venerable, rational, selfish maximizer of traditional economic theory; *Homo socialis* is the other-regarding agent who cares about fairness, reciprocity, and the well-being of others; and *Homo moralis* is the Aristotelian bearer of nonconsequentialist character virtues. The new types of public *persona* are *Homo autisticus* who behaves politically just as *Homo economicus* does privately, whereas *Homo parochialis* votes and engages in collective action on behalf of the narrow interests of the demographic, ethnic, and/or social status groups with which he identifies. Finally, *Homo universalis* acts politically to achieve what he considers the best state for the larger society—for instance, reflecting John Rawls's (1971) *veil of ignorance*, John Harsanyi's (1977) *criterion of universality*, or John Roemer's (2010) *Kantian equilibrium*.

Probably *Homo autisticus* does not exist because a self-regarding agent will never do anything except for its consequences, and the concept of rule-consequentialism is difficult to comprehend when one is the only "like-minded" person. Interestingly, the individual whose private persona is social is generally considered altruistic, whereas the individual whose public persona is social is often considered selfish, acting in a partisan manner on behalf of the narrow interests of the social networks to which he belongs. Of course *Homo parochialis* is altruistic toward these social networks.

Table 3.1 A Typology of Human Motivations

Persona	Personal	Social	Universal
Private	*Homo* economics	*Homo* socialis	*Homo* vertus
Public	*Homo* autisticus	*Homo* parochialis	*Homo* universalis

The Evolutionary Emergence of Private Morality

By *cooperation* we mean engaging with others in a mutually beneficial activity. Cooperative behavior may confer net benefits on the individual cooperator, and thus can be motivated entirely by self-interest. In this case, cooperation is a form of *mutualism*. Cooperation may also be a net cost to the individual, but the benefits may accrue to a close relative. We call this *kin altruism*. Cooperation can additionally take the form of one individual's costly contribution to the welfare of another individual being reliably reciprocated at a future date. This is often called *reciprocal altruism* (Trivers 1971), although it is really just tit-for-tat mutualism. However, important forms of cooperation impose net costs upon individuals, the beneficiaries may not be close kin, and the benefit to others may not be expected to be repaid in the future. This cooperative behavior is true altruism.

The evolution of mutualistic cooperation and kin altruism is easily explained. Cooperation among close family members evolves by natural selection because the benefits of cooperative actions are conferred on the close genetic relatives of the cooperator, thereby helping to proliferate genes associated with the cooperative behavior. Kin altruism and mutualism explain many forms of human cooperation, particularly those occurring in families or in frequently repeated two-person interactions. But, these models fail to explain two facts about human cooperation: that it takes place in groups far larger than the immediate family, and that both in real life and in laboratory experiments, it occurs in interactions that are unlikely to be repeated, and in which it is impossible to obtain reputational gains from cooperating. These forms of behavior are regulated by moral sentiments.

The most parsimonious proximal explanation of altruistic cooperation, one that is supported by extensive experimental and everyday-life evidence, is that people gain pleasure from cooperating and feel morally obligated to cooperate with like-minded people. People also enjoy punishing those who exploit the cooperation of others. Free-riders frequently feel guilty, and if they are sanctioned by others, they may feel ashamed. We term these feelings *social preferences*. Social preferences include a concern, positive or negative, for the well-being of others, as well as a desire to uphold ethical norms.

The Roots of Social Preferences

Why are the social preferences that sustain altruistic cooperation in daily life so common? Early human environments are part of the answer. Our Late Pleistocene ancestors inhabited the large-mammal-rich African savannah and other environments in which cooperation in acquiring and sharing food yielded substantial benefits at relatively low cost. The slow human life history with prolonged periods of dependency of the young also made the cooperation of nonkin in child rearing and provisioning beneficial. As a result, members of groups that sustained cooperative strategies for provisioning, child rearing, sanctioning noncooperators, defending against hostile neighbors, and sharing information truthfully had significant advantages over members of noncooperative groups.

There are several reasons why these altruistic social preferences supporting cooperation outcompeted amoral self-interest. First, human groups devised ways to protect their altruistic members from exploitation by the self-regarding. Prominent among these is the *collective punishment of miscreants* (Boyd et al. 2010), including the public-spirited shunning, ostracism, and even execution of free-riders and others who violate cooperative norms.

Second, humans adopted elaborate systems of *socialization* that led individuals to internalize the norms that induce cooperation, so that contributing to common projects and punishing defectors became objectives in their own right rather than constraints on behavior. Together, the internalization of norms and the protection of the altruists from exploitation served to offset, at least in part, the competitive handicaps born by those who were motivated to bear personal costs to benefit others.

Third, between-group competition for resources and survival was and remains a decisive force in human evolutionary dynamics. Groups with many cooperative members tended to survive these challenges and to encroach on the territory of the less cooperative groups, thereby gaining both reproductive advantages and proliferating cooperative behaviors through cultural transmission. The extraordinarily high stakes of intergroup competition and the contribution of altruistic cooperators to success in these contests meant that sacrifice on behalf of others, extending beyond the immediate family and even to virtual strangers, could proliferate (Choi and Bowles 2007; Bowles 2009).

This is part of the reason why humans became extraordinarily group-minded, favoring cooperation with insiders and often expressing hostility toward outsiders. Boundary maintenance supported within-group cooperation and exchange by limiting group size and within-group linguistic, normative, and other forms of heterogeneity. Insider favoritism also sustained the between-group conflicts and differences in behavior that made group competition a powerful evolutionary force.

In short, we became a cooperative species because cooperation was highly beneficial to the members of groups that practiced it, and we were able to construct social institutions that minimized the disadvantages of those with social preferences in competition with self-regarding fellow group members while heightening the group-level advantages associated with the high levels of cooperation that these social preferences allowed. These institutions proliferated because the groups that adopted them secured high levels of within-group cooperation, which in turn favored the groups' survival as a biological and cultural entity in the face of environmental, military, and other challenges.

The Evolutionary Emergence of the Public Persona

Nonhuman species, even if highly social, do not engage in activities that structure the social rules that regulate their lives. Therefore, there are no politics and no public sphere in these species, and hence its members have no public persona. How, then, might a public persona have arisen in the hominin line leading up to *Homo sapiens*?

In a related paper, Carel van Schaik, Christopher Boehm, and I (Gintis et al. 2015) supply an answer grounded in the information available to us from a variety of fields, including paleontology, primatology, the anthropology of contemporary hunter-gatherer groups, animal behavior theory, and genetics. We propose that the emergence of bipedalism, cooperative breeding, and lethal weapons (stones and wooden spears) in the hominin line, together with favorable climate change, made the collaborative hunting and scavenging of large game fitness enhancing. Lethal weapons are the most unique of these innovations, for other predators, such as lions, tigers and other big cats, wolves, foxes, and other canines, use only their natural weapons—sharp claws and teeth, powerful jaws, and great speed—in hunting, whereas none of these endowments was available to early hominins. Lethal hunting weapons, moreover, transformed human sociopolitical life because they could be applied to humans just as easily as to other animals.

The combination of the need for collaboration and the availability of lethal weapons in early hominin society undermined the social dominance hierarchy characteristic of primate and earlier hominin groups, which was based on pure physical prowess. The successful sociopolitical structure that ultimately replaced the ancestral social dominance hierarchy was an egalitarian political system in which lethal weapons made possible group control of leaders, and group success depended on the ability of leaders to persuade and motivate and of followers to contribute to a consensual decision process. The heightened social value of non-authoritarian leadership entailed enhanced biological fitness for leadership traits such as linguistic facility, ability to form and influence coalitions, and, indeed, for hypercognition in general.

This egalitarian political system persisted until cultural changes in the Holocene fostered the accumulation of material wealth, through which it became possible once again to sustain a social dominance hierarchy with strong authoritarian leaders.

Conclusion

This chapter provided evidence for a model of human behavior based on the rational actor model, in which individuals have both a private persona and a public persona, and their preferences range over personal, social, and universal modes of our private personas and in most of our activities in the public sphere. Morality in this model is defined in behavioral terms: moral choices are those made in social and universalist modes. The public sphere in this model is an arena in which preferences and actions are primarily nonconsequentialist. The other-regarding preferences of *Homo socialis* and the character virtues of *Homo vertus* are underpinnings of civil society, whereas *Homo parochialis* and *Homo universalis* make possible the varieties of political life characteristic of our species.

References

Abbott, Richard J., Juliet K. James, Richard I. Milne, and Amanda C. M. Gillies. 2003. Plant introductions, hybridization and gene flow. *Philosophical Transactions of the Royal Society of London B* 358: 1123–1132.

Alexander, Richard D. 1987. The biology of moral systems. New York: Aldine.

Allman, John, Atiya Hakeem, and Karli Watson. 2002. Two phylogenetic specializations in the human brain. *Neuroscientist* 8: 335–346.

Aristotle. 2002 [350BC]. *Nicomachean ethics*. Newburyport, MA: Focus Publishing.

Beer, Jennifer S., Erin A. Heerey, Dacher Keltner, Donatella Skabini, and Robert T. Knight. 2003. The regulatory function of self-conscious emotion: Insights from patients with orbitofrontal damage. *Journal of Personality and Social Psychology* 65: 594–604.

Belin, Pascal, Robert J. Zatorre, Phillipe Lafaille, Pierre Ahad, and Bruce Pike. 2000. Voice-selective areas in human auditory cortex. *Nature* 403: 309–312.

Berg, Joyce E., John W. Dickhaut, and Kevin McCabe. 1995. Trust, reciprocity, and social history. *Games and Economic Behavior* 10: 122–142.

Binder, Jeffrey, Julie A. Frost, Thomas A. Hammeke, Robert W. Cox, Stephen M. Rao, and Thomas Prieto. 1997. Human brain language areas identified by functional magnetic resonance imaging. *Journal of Neuroscience* 17: 353–362.

Blount, Sally. 1995. When social outcomes aren't fair: The effect of causal attributions on preferences. *Organizational Behavior and Human Decision Processes* 63: 131–144.

Boehm, Christopher 1999. *Hierarchy in the forest: The evolution of egalitarian behavior*. Cambridge, MA: Harvard University Press.

Boles, Terry L., Rachel T. A. Croson, and J. Keith Murnighan. 2000. Deception and retribution in repeated ultimatum bargaining. *Organizational Behavior and Human Decision Processes* 83: 235–259.

Bolton, Gary E., and Rami Zwick. 1995. Anonymity versus punishment in ultimatum games. *Games and Economic Behavior* 10: 95–121.

Bonner, John T. 1984. *The evolution of culture in animals.* Princeton: Princeton University Press.

Bowles, Samuel 2009. Did warfare among ancestral hunter-gatherer groups affect the evolution of human social behaviors? *Science* 324: 1293–1298.

Bowles, Samuel, and Herbert Gintis. 1986. *Democracy and capitalism: Property, community, and the contradictions of modern social thought.* New York: Basic Books.

———. 2011. *A cooperative species: Human reciprocity and its evolution.* Princeton: Princeton University Press.

Boyd, Robert, Herbert Gintis, and Samuel Bowles. 2010. Coordinated punishment of defectors sustains cooperation and can proliferate when rare. *Science* 328: 617–620.

Boyd, Robert, and Peter J. Richerson. 1985. *Culture and the evolutionary process.* Chicago: University of Chicago Press.

———. 1988. An evolutionary model of social learning: The effects of spatial and temporal variation. In: T. R. Zentall and G. Galef, Jr., eds., *Social learning: Psychological and biological perspectives,* 29–48. Hillsdale, NY: Erlbaum.

Brown, James H., and Mark V. Lomolino. 1998. *Biogeography.* Sunderland, MA: Sinauer.

Burks, Stephen V., Jeffrey P. Carpenter, and Eric Verhoogen. 2003. Playing both roles in the trust game. *Journal of Economic Behavior and Organization* 51: 195–216.

Burrows, Anne M. 2008. The facial expression musculature in primates and its evolutionary significance. *BioEssays* 30: 212–225.

Camerer, Colin, and Richard H. Thaler. 1995. Ultimatums, dictators, and manners. *Journal of Economic Perspectives* 9: 209–219.

Camille, Natalie, Giorgio Coricelli, Jerome Sallet, Pascale Pradat-Diehl, Jean-René Duhamel, and Angela Sirigu. 2004. The involvement of the orbitofrontal cortex in the experience of regret. *Science* 304: 1167–1170.

Cavalli-Sforza, Luigi L., and Marcus W. Feldman. 1981. *Cultural transmission and evolution.* Princeton: Princeton University Press.

———. 1982. Theory and observation in cultural transmission. *Science* 218: 19–27.

Charness, Gary, and Martin Dufwenberg. 2006. Promises and partnership. *Econometrica* 74: 1579–1601.

Choi, Jung-Kyoo., and Samuel Bowles. 2007. The coevolution of parochial altruism and war. *Science* 318: 636–640.

Dawkins, Richard 1976. *The selfish gene.* Oxford: Oxford University Press.

———. 1982. *The extended phenotype: The gene as the unit of selection.* Oxford: Freeman.

Downs, Anthony 1957. *An economic theory of democracy.* Boston: Harper and Row.

Dunbar, Robin M. 1993. Coevolution of neocortical size, group size and language in humans. *Behavioral and Brain Sciences* 16: 681–735.

———. 2005. *The human story.* New York: Faber and Faber.

Edgeworth, Francis Y. 1881. *Mathematical psychics: An essay on the application of mathematics to the moral sciences.* London: Kegan Paul.

Edlin, Aaron, Andrew Gelman, and Noah Kaplan. 2007. Voting as a rational choice: Why and how people vote to improve the well-being of others. *Rationality and Society* 19: 293–314.

Fedderson, Timothy, and Alvaro Sandroni. 2006. A theory of participation in elections. *American Economic Review* 96: 1271–1282.

Feldman, Marcus W., and Lev A. Zhivotovsky. 1992. Gene-culture coevolution: Toward a general theory of vertical transmission. *Proceedings of the National Academy of Sciences of the United States of America* 89: 11935–11938.

Fudenberg, Drew, David K. Levine, and Eric Maskin. 1994. The folk theorem with imperfect public information. *Econometrica* 62: 997–1039.

Gadagkar, Raghavendra 1991. On testing the role of genetic asymmetries created by haplodiploidy in the evolution of eusociality in the hymenoptera. *Journal of Genetics* 70: 1–31.

Ghiselin, Michael T. 1974. *The economy of nature and the evolution of sex.* Berkeley: University of California Press.

Gigerenzer, Gerd, and Peter M. Todd. 1999. *Simple heuristics that make us smart.* New York: Oxford University Press.

Gintis, Herbert 2009. *The bounds of reason: Game theory and the unification of the behavioral sciences.* Princeton: Princeton University Press.

———. 2014. Inclusive fitness and the sociobiology of the genome. *Biology and Philosophy* 29: 477–515.

Gintis, Herbert, and Dirk Helbing. 2015. *Homo socialis*: An analytical core for sociological theory. *Review of Behavioral Economics* 2: 1–59.

Gintis, Herbert, Carel van Schaik, and Christopher Boehm. 2015. Zoon politikon: The evolutionary origins of human political systems. *Current Anthropology* 56: 327–353.

Giugni, Marco G., Doug McAdam, and Charles Tilly. 1998. *From contention to democracy.* London: Rowman and Littlefield.

Gneezy, Uri 2005. Deception: The role of consequences. *American Economic Review* 95: 384–394.

Greenberg, Martin S., and David M. Frisch. 1972. Effect of intentionality on willingness to reciprocate a favor. *Journal of Experimental Social Psychology* 8: 99–111.

Gunnthorsdottir, Anna, Kevin McCabe, and Vernon L. Smith. 2002. Using the Machiavellianism instrument to predict trustworthiness in a bargaining game. *Journal of Economic Psychology* 23: 49–66.

Güth, W., R. Schmittberger, and B. Schwarze. 1982. An experimental analysis of ultimatum bargaining. *Journal of Economic Behavior and Organization* 3: 367–388.

Güth, Werner, and Reinhard Tietz. 1990. Ultimatum bargaining behavior: A survey and comparison of experimental results. *Journal of Economic Psychology* 11: 417–449.

Haidt, Jonathan 2012. *The righteous mind: Why good people are divided by politics and religion*. New York: Pantheon.

Hamlin, Alan, and Colin Jennings. 2011. Expressive political behavior: Foundations, scope and implications. *British Journal of Political Science* 41: 645–670.

Harsanyi, John C. 1977. Morality and the theory of rational behavior. *Social Research* 44: 623–656.

Henrich, Joseph, Robert Boyd, Samuel Bowles, Colin Camerer, Ernst Fehr, and Herbert Gintis. 2004. *Foundations of human sociality: Economic experiments and ethnographic evidence from fifteen small-scale societies.* Oxford: Oxford University Press.

Henrich, Joseph, and Francisco Gil-White. 2001. The evolution of prestige: Freely conferred status as a mechanism for enhancing the benefits of cultural transmission. *Evolution and Human Behavior* 22: 165–196.

Hobbes, Thomas 1968 [1651]. *Leviathan.* Ed., C. B. MacPherson. New York: Penguin.

Holden, Clare J. 2002. Bantu language trees reflect the spread of farming across sub-Saharan Africa: A maximum-parsimony analysis. *Proceedings of the Royal Society of London B* 269: 793–799.

Holden, Clare J., and Ruth Mace. 2003. Spread of cattle led to the loss of matrilineal descent in Africa: A coevolutionary analysis. *Proceedings of the Royal Society of London B* 270: 2425–2433.

Hooker, Brad 2015. Rule consequentialism. *The Stanford Encyclopedia of Philosophy* (Winter 2015 Edition), Edward N. Zalta (ed.), URL = <http://plato.stanford.edu/archives/win2015/entries/consequentialism-rule/>.

Huxley, Julian S. 1955. Evolution, cultural and biological. *Yearbook of Anthropology* 2–25.

Ihara, Yasuo 2011. Evolution of culture-dependent discriminate sociality: A gene-culture coevolutionary model. *Proceedings of the Royal Society of London B* 366: 889–900.

Jablonka, Eva, and Marion J. Lamb. 1995. *Epigenetic inheritance and evolution: The Lamarckian case.* Oxford: Oxford University Press.

James, William 1880. Great men, great thoughts, and the environment. *Atlantic Monthly* 46: 441–459.

Jurmain, Robert, H. Nelson, Lynn Kilgore, and Wenda Travathan. 1997. *Introduction to physical anthropology*. Cincinnati: Wadsworth.

Kelly, Robert L. 1995. *The foraging spectrum: Diversity in hunter-gatherer lifeways.* Washington, DC: The Smithsonian Institute.

Mace, Ruth, and Mark Pagel. 1994. The comparative method in anthropology. *Current Anthropology* 35: 549–564.

Mednick, Sarnoff A., Lis Kirkegaard-Sorenson, Barry Hutchings, Joachim Knop, Raben Rosenberg, and Fini Schulsinger. 1977. An example of bio-social interaction research: The interplay of socio-environmental and individual factors in the etiology of criminal behavior. In Sarnoff A. Mednick and Karl O. Christiansen, eds., *Biosocial bases of criminal behavior,* 9–24. New York: Gardner.

Mesoudi, Alex, Andrew Whiten, and Kevin N. Laland. 2006. Towards a unified science of cultural evolution. *Behavioral and Brain Sciences* 29: 329–383.

Miller, Bruce L., Adam Darby, D. Frank Benson, Jeffrey L. Cummings, and Michael H. Miller. 1997. Aggressive, socially disruptive and antisocial behaviour associated with frontotemporal dementia. *British Journal of Psychiatry* 170: 150–154.

Moll, Jorge, Roland Zahn, Ricardo di Oliveira-Souza, Frank Krueger, and Jordan Grafman. 2005. The neural basis of human moral cognition. *Nature Neuroscience* 6: 799–809.

Newman, Mark, Albert-László. Barabasi, and Duncan J. Watts. 2006. *The structure and dynamics of networks.* Princeton: Princeton University Press.

O'Brien, Michael J., and R. Lee Lyman. 2000. *Applying evolutionary archaeology.* New York: Kluwer Academic.

Odling-Smee, F. John, Kevin N. Laland, and Marcus W. Feldman. 2003. *Niche construction: The neglected process in evolution.* Princeton: Princeton University Press.

Olson, Mancur 1965. *The logic of collective action: Public goods and the theory of groups.* Cambridge: Harvard University Press.

Parsons, Talcott 1964. Evolutionary universals in society. *American Sociological Review* 29: 339–357.

Pinker, Steven. 2002. *The blank slate: The modern denial of human nature.* New York: Viking.

Popper, Karl, 1979. *Objective knowledge: An evolutionary approach.* Oxford: Clarendon Press.

Rawls, John 1971. *A theory of justice.* Cambridge, MA: Harvard University Press.

Relethford, John H. 2007. *The human species: An introduction to biological anthropology.* New York: McGraw-Hill.

Richerson, Peter J., and Robert Boyd. 1998. The evolution of ultrasociality. In Irenäus Eibl-Eibesfeldt and Frank K. Salter, eds., *Indoctrinability, ideology and warfare,* 71–96. New York: Berghahn Books.

———. 2004. *Not by genes alone.* Chicago: University of Chicago Press.

Riker, William H., and Peter C. Ordeshook. 1968. A theory of the calculus of voting. *American Political Science Review* 62: 25–42.

Rivera, Maria C., and James A. Lake. 2004. The ring of life provides evidence for a genome fusion origin of eukaryotes. *Nature* 431: 152–155.

Roemer, John. 2010. Kantian equilibrium. *Scandinavian Journal of Economics* 112: 1–24.

Roth, Alvin E., Vesna Prasnikar, Masahiro Okuno-Fujiwara, and Shmuel Zamir. 1991. Bargaining and market behavior in Jerusalem, Ljubljana, Pittsburgh, and Tokyo: An experimental study. *American Economic Review* 81: 1068–1095.

Savage, Leonard J. 1954. *The foundations of statistics.* New York: John Wiley and Sons.

Schulkin, Jay. 2000. *Roots of social sensitivity and neural function*. Cambridge, MA: MIT Press.

Seeley, Thomas D. 1997. Honey bee colonies are group-level adaptive units. *American Naturalist* 150: S22–S41.

Shennan, Stephen. 1997. *Quantifying archaeology*. Edinburgh: Edinburgh University Press.

Skibo, James M., and R. Alexander Bentley. 2003. *Complex systems and archaeology*. Salt Lake City: University of Utah Press.

Sterelny, Kim 2011. From hominins to humans: How *Sapiens* became behaviourally modern. *Proceedings of the Royal Society of London B* 366: 809–822.

Tooby, John, and Leda Cosmides. 1992. The psychological foundations of culture. In J. H. Barkow, L. Cosmides, and J. Tooby, eds., *The adapted mind: Evolutionary psychology and the generation of culture*, 19–136. New York. Oxford University Press.

Trivers, Robert L. 1971. The evolution of reciprocal altruism. *Quarterly Review of Biology* 46: 35–57.

Wilson, Edward O., and Bert Hölldobler. 2005. Eusociality: Origin and consequences. *Proceedings of the National Academy of Sciences of the United States of America* 102: 13367–13371.

Winterhalder, Bruce, and Eric A. Smith. 1992. *Evolutionary ecology and human behavior*. New York: Aldine de Gruyter.

Zajonc, Robert B. 1980. Feeling and thinking: Preferences need no inferences. *American Psychologist* 35: 151–175.

4 HUMAN KINSHIP AS A GREEN BEARD

Henry Harpending[†] and Nathan Harris

Humans have a lot of difficulty with ethnic diversity. Some have speculated that ethnic discomforts are somehow based on or derived from genetic differences. In raw form, the question arises whether "racism is in our genes." A recent article in the *Atlantic* by Robert Wright gives assurances that it could not be (Wright 2012). Wright's assurances are based on a widespread but not well-supported argument about the human past—that we never encountered people significantly different from us until the past few thousand years. Consequently, the ecological conditions that could have favored ethnic discrimination never occurred in our evolution. That argument demands a particular ecology and demography in our past, and it demands that evolution takes place over very long time periods, hundreds of millennia or longer.

There are grounds to doubt both fundamental premises of the argument. One premise is that, in our past (whatever period that is supposed to occupy), humans were essentially a uniform carpet of sessile groups with substantial ongoing gene exchange with neighboring groups. Another premise is that the relevant past is tens to hundreds of thousands of millennia.

We know there are cases in which genetically different ethnic groups experience no apparent discomfort with each other, for instance the Parsi in India. Other Indians regard the Parsi as a national resource (Nelson 2012). There are also familiar cases of ethnic antagonism between genetically identical groups, for instance Mormons and non-Mormons in Utah.

The notion of "in our genes" deserves some attention, especially because "genetic" is often contrasted with "learned" or "environmental." We all learned our spoken language; many, but not all, of us are reflexively afraid of snakes; and many of us learned calculus.

Language never had to be taught because humans are "prepared" to learn language. Many argue we are also prepared to fear snakes, but calculus is an entirely different matter. We had to sit and think hard about it, and spend hours doing exercises that molded what understanding we gained.

The issue in this chapter is whether we are easily taught ethnic antagonism. Is it like language, so easily learned that it never need be taught? Or is it like calculus, which must always be taught with great effort? Or is it somewhere in between?

We could presumably answer this question with a properly designed experiment, but that is never going to occur. We all correctly regard experimentation like that with humans as unethical. An alternate approach, used in most of the sciences, is to make a model and see what the model predicts. A good model should tell us whether we ought to have ethnic antagonism built-in—in other words, easily learned, or never built-in, like calculus. It ought to tell us the conditions under which one or the other might have occurred in our past and, best of all, it ought to suggest to us ways to falsify the predictions of the model.

Genetic Background

W. D. Hamilton (1964a; 1964b) created a revolution in behavioral ecology with a pair of papers that defined and clarified what today is called *indirect* or *extended fitness*. If I take care of my child, I not only augment that child's Darwinian fitness but also my own, in proportion to the amount of my genome shared with that child. The early version of Hamilton's theory was described in terms of the "coefficient of relationship," the fraction of shared genes between an actor and the recipient. In the case of my child, this is conventionally thought to be one-half.

Although many were immediately enthusiastic about the application of this theory when it was first published, especially entomologists, who deal with strange variants of sexual reproduction, others thought it was trivial and obvious, so they paid little attention. A colleague remarked to me during the 1970s, that "we don't need inclusive fitness theory to understand why parental care evolved." The research of Daly and Wilson (1988) helped inclusive theory gain traction and respect in anthropology and the social sciences. They demonstrated that children living in households with adult males who were not their biological fathers were at much greater risk of severe mistreatment and even murder.

It soon became apparent that "fraction of shared genes" is a slippery idea. Modern population genetics bypasses the old distinction between "identity by descent" and "identity in state" because evolution acts on the products of genes, and identical genes are simply identical or not, regardless of the origin of their identity. It is true enough that I share half my autosomal genes with my child, but if I am inbred, then I share slightly more than one-half because the inbreeding means that there are correlations between the alleles at any locus in me. Hamilton himself came to appreciate this and, as a result, extended his ideas beyond close family relationships to more distant relationships—for example, within ethnic groups in a multiethnic context (Hamilton 1975).

For diploid organisms, like humans, the difficulties with relationship are easily resolved by reasoning about the coefficient of kinship between two organisms (Bulmer 1994) defined like this: Pick a random allele from a locus in individual A, pick a random allele from the same locus in individual B, and ask whether they are identical. Then, ask what the frequency of the allele is in the population as a whole, and call that p. Then, the probability that the gene from B is identical to the gene chosen from A is

$$p_B = p_A F_{AB} + p(1 - F_A B)$$

leading to an estimate for F from this locus as

$$F_{AB} = (p_B - p) / (1 - p)$$

This estimate can have bizarre properties. For some possible genotype configurations it is nearly unbounded above 1 and below –1, whereas the "real" value of F_{AB} must be between +1 and –1. On the other hand, averaged over hundreds of thousands of loci, it returns correct results. Averaging, the kinship between person A and person B at a locus is computed as

$$(p_A - p)(p_B - p) / p(1 - p)$$

where p_A and p_B are the frequencies in the two individuals, with possible values 1, 0.5, and 0, and p is the population average (Yang et al. 2010; Davies et al. 2011). With many loci, hundreds of thousands in the current case, one sums the numerators of the previous expression of loci, sums the denominator, and computes the ratio. Yang et al. (2010) and Davies et al. (2011) give the computational formulae.

Genomics has provided a powerful tool for assessing genetic kinship, especially kinship between people who do not know each other. An unfortunate downside of the new technology is semantic confusion. In this chapter, I use *kinship* to mean shared genes whereas many think of kinship as a genealogical or social relationship. The two (or more) meanings of kinship are overlapping, but they are different enough in practice that caution is in order.

We have used single nucleotide polymorphism genotype data from the Human Genome Diversity Project (HGDP) project (Li et al. 2008) to examine kinship between individuals in several populations from the database. The HGDP was a loosely coordinated project to collect genetic samples from several dozen human populations around the world. In each population, a few dozen or more individuals were sampled. Samples were included from large cosmopolitan populations, such as the French, Japanese, and Mormons (from Utah), as well as from small, isolated tribal populations like the Surui and Karitiana of the Amazon Basin. Individual populations samples are of more than 20 and fewer than 50 individuals. There are several recently admixed populations: a sample of Uyghurs from central Asia and a sample of African Americans from the U.S. Southwest.

Pairwise kinship between two individuals immediately informs about the genetic "interest" that one individual should share with the other. If the social and economic system is such that individuals can reward (or harm) each other according to their genetic interests, then the prediction is that dispositions for discrimination and nepotism, according to shared kinship, could have evolved, such that discrimination and nepotism would be easily learned or "natural." Two important questions to ask are (a) whether there are enough differences in kinship between individuals of unknown genealogical relationship in populations to have favored cryptic kinship recognition by natural selection and (b) whether we have mechanisms available to us that could detect cryptic kinship.

Cultural Background

Since the 1970s, many evolutionary social scientists have believed that technologically primitive groups, especially foragers, offer clues to our species' proclivities and potentials. The reasoning was that, because we were hunter-gatherers for 99% of our evolution, we must have been somehow "shaped" by evolution to be like them. Some have called this the search for "Mr. Natural," whereas others refer to this hypothetical ancestral state as the "environment of evolutionary adaptedness," or EEA. Although the rationale is subject to criticism, the search produced a lot of useful and interesting ethnography, ethnography addressing evolutionary rather than simply social questions. Where does food come from? How does reproductive competition work itself out? What determines the fitness of individuals?

In retrospect, that are several weaknesses with EEA theory. First, what does "99% of our existence" mean? If distinctly modern humans appeared at least 40,000 years ago, and if many of our ancestors have practiced agriculture for 10,000 years, then they were foragers for 75% of our existence, not 99%. The

99% argument requires that we are living relics of life as *Homo erectus* or even some flavor of australopithecines, for the past million years.

Second, EEA theory needs there to have been some identifiable EEA, but the archaeological and ecological record of the past million years shows much diversity, and its commonalities fail to answer the most basic questions about this social environment. For example, were males in the EEA paternal males who provisioned their own offspring? Were they instead competitive, violence-prone gaudy males who provided little in the way of subsistence for women and children? In Europe, in the several millennia before the last glacial maximum, we find upper Paleolithic remains with many signatures of the latter, and Mesolithic remains looking like the former.

Two candidates for our EEA, for "Mr. Natural," emerged in the textbooks of the 1970s. One was !Kung Bushmen of the northern Kalahari in Africa (Lee and DeVore 1976), and the other was the Yanomamo of the Amazon Basin (Chagnon 1968). They could hardly have been more different.

The Bushmen were laid-back, relaxed people, seemingly tailor-made for the popular culture of the time—Kalahari hippies with beads and sandals even. The males were not belligerent and not loud, and they put in a lot of time in resource acquisition that was distributed not only to their immediate families, but also to every member of the local band. Women gathered and provisioned their own children and husbands with the harvest. Nuclear families were durable. Polygyny was certainly possible, but it was ordinarily not real reproductive polygyny; it was more likely an older widow living with her sister and her brother-in-law as a "co-wife." Marital distances (distances between birthplaces of husbands and wives) were large, so the population was outbred. Groups separated by hundreds of kilometers displayed only small genetic differences. Population was stationary, a significant finding in that era of concern about world population. They were no threat to the environment. Settlements, called "bands," were not persistent social units, but families were. Yellen and Harpending (1972) show maps of where families of one band had lived in the past year, with no apparent correlation between one family and the next.

The Yanomamo were in many ways opposite to the !Kung. They were engaged in chronic local warfare and raiding, so it seems no one was very relaxed. Pair bonds were unstable and males were not focused very much on provisioning. With their transient gardens (swiddens), they were a real threat to the environment because abandoned swiddens in tropical environments take a long time to recover. The rate of reproduction was so high that the population was approximately doubling each generation. The Yanomamo were predatory expansionists with gaudy, loud, violent males. There was a substantial payoff to male violence. Males who had killed someone enjoyed more reproductive success than males who had never killed anyone (Chagnon 1988). Marital distances were small. Villages were persistent clusters of related people. A consequence is that villages were genetically distinct from each other, even with the small set of genetic markers available at the time.

These two models for our EEA provide dramatically different models of the social contexts for the evolution of ethnic antagonism and kin nepotism. In a Bushman world, with its stationary populations and high outbreeding, there is no ethnic differentiation. Two people separated by several hundred kilometers are scarcely more different from each other than two people from the same settlement. In a Yanomamo world, on the other hand, the landscape is one of kaleidoscopic expansions and extinctions. In a lifetime, people from one group would ordinarily have violent interactions with people from very different groups. In their studies of the Upper Midwest before European contact, George Milner et al. (1991) provided an excellent archaeological example of this world. By the time of Columbus, struggles with invading Oneota had essentially depopulated the upper Midwest. This paper was a shock to many anthropologists at the time, who enjoyed a stereotype that low-technology people were uniformly gentle and cooperative. Two decades later, such findings are mainstream.

Expansions, disruptions, and extinctions are familiar to us from history: the European invasion of the New World, the Bantu and Han expansions, the Indo-Europeans, the Mongols, and others. We have no good reason to think that the same pattern did not occur further back in human history. Patterns are provided by the Oneota case, the Numic expansion in the American West, the arrival of Navaho and Apache in the American Southwest, and the expansion of Inuit at the expense of the Dorset in the arctic.

Evidence from the Genome

As mentioned, the HGDP contains several samples from large cosmopolitan populations such as Mormons from Utah, French, Japanese, and two Chinese samples; and other samples from small, isolated populations and from ethnic groups within larger political units. Details of the samples are not available because of concerns about protecting the privacy of those who furnished the samples. We are forced to assume the samples are not biased in any obvious ways, but we will see that the occasional pairs of relatives do turn up in them. To determine the extent of cryptic kinship, and thus to see whether natural selection could favor recognition of kin, we can simply look at histograms of kinship between all possible pairs of individuals in the HGDP samples. We first consider pairwise kinship in the samples of French and Japanese.

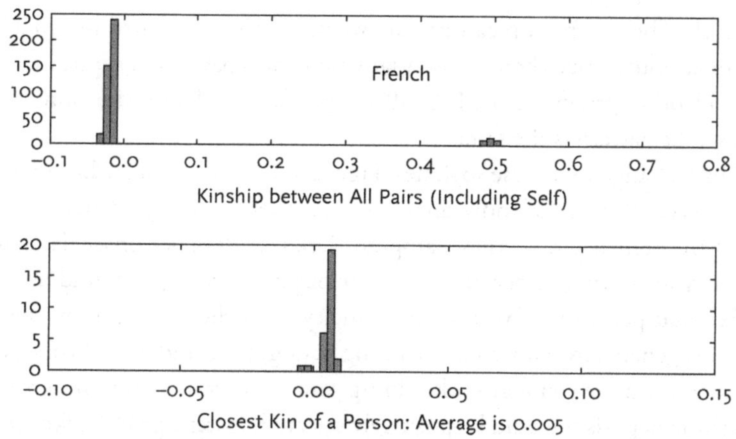

FIGURE 4.1 Kinship among French. The top panel is a histogram of kinship between all pairs of individuals, including kinship with self, shown by the small number of comparison at F ≈ 0.5. The bottom panel shows, for each person in the sample, kinship with his closest kin in the sample. Because overall mean pairwise kinship must be zero, and because kinship with self around F = 0.5 dominates the average in small samples such as this, kinship in the bottom panel has been adjusted to a mean of zero.

Recalling that kinship with self is roughly 0.5, so that kinship with a child is 0.25, a grandchild 0.125, a grandchild's grandchild is 0.006, and so on, it is immediately apparent from Figures 4.1 and 4.2 that there is not enough diversity in kinship with other members of the same group to have ever favored cryptic kin recognition or spontaneous nepotistic groups. In these large cosmopolitan

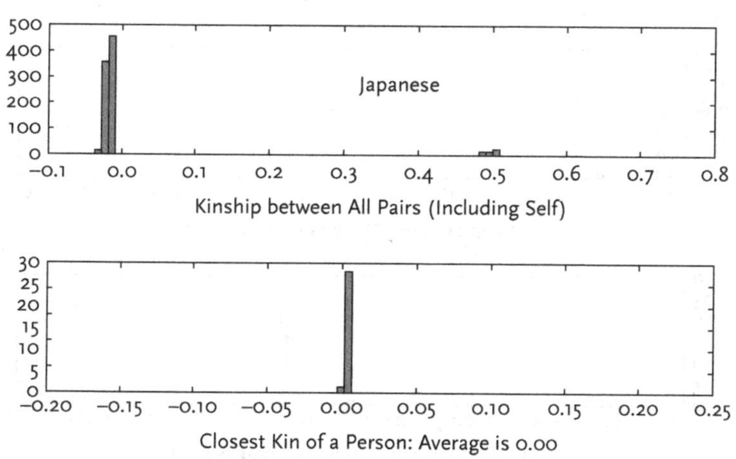

FIGURE 4.2 Kinship among Japanese. See Figure 4.1 for details about the panels.

groups, the "best" a person can do (or "worst" if one is to harm someone) is so small in absolute value there is essentially no fitness payoff to cryptic kin recognition. In other groups in the HGDP sample, such as Utah Mormons or Han Chinese, the pattern is the same.

Things change in diverse societies. Figure 4.3 shows the distribution of kinship between all pairs of individuals in a (synthetic) society created by simply pooling the French and Japanese samples. Here a random person is related to you either as your great-grandchild or as your negative great-grandchild according whether that person is of your own ethnicity or of the other. Aiding someone like you is genetically equivalent to aiding your great-grandchild. Worse, harming someone unlike you is also like aiding your great-grandchild. In a group like this, ethnic nepotism would be positively selected, fostering all the familiar difficulties humans have with diversity.

The picture so far is that in large homogeneous societies, kin recognition would have no fitness payoff; in diverse societies, the fitness payoff to nepotism and ethnocentrism would be immediate and substantial. Is this relevant to understanding our species' propensity for ethnic antagonism? A possibility is that ethnic antagonism is relatively new and that it is an evolved response to life in ancient cities, because large urban centers arising in the context of wide trade networks could easily have generated ethnic mixes that mirror our synthetic mixture of French and Japanese.

If long-range movements did, in the past, generate societies such as our synthetic mixture, how long did the diversity persist? There are several relevant populations to examine in the HGDP sample. First, Uyghurs are a central Asian group descended from Tocharians, the easternmost extension of Indo-European

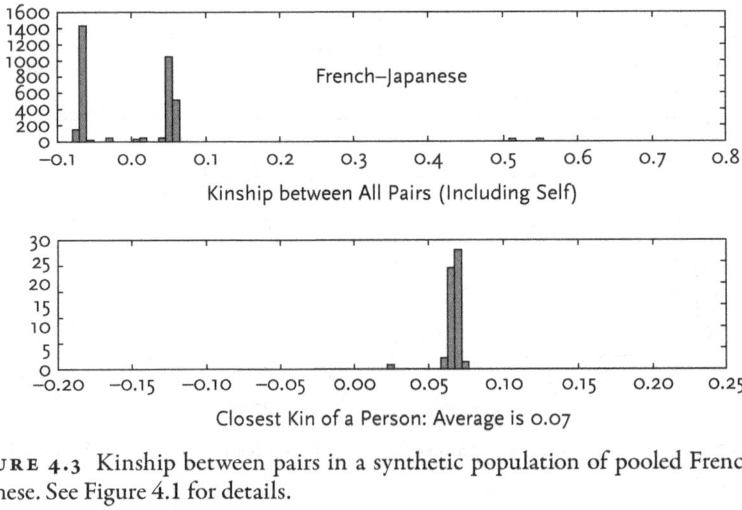

FIGURE 4.3 Kinship between pairs in a synthetic population of pooled French and Japanese. See Figure 4.1 for details.

speakers in the Tarim Basin of northwest China. They are, today, roughly half descended from these Europeans and half from East Asian peoples. The second population of interest is a group of African Americans from the U.S. Southwest.

Figure 4.4 shows the estimated ancestry of individuals in each of these populations. (This figure is reproduced with permission from a blog [Khan 2010]: http://blogs.discovermagazine.com/gnxp/?p=7643#.UNi09Ylbzt8.) In each part of the figure, individuals are represented by horizontal lines, the colors of which show the ancestral contributions to that individual. The Uyghurs have been mixed for a millennium or so. They are an old admixed population, old enough that ancestries of individuals have been homogenized. Every individual is around half European and half East Asian. What this means is that they have a "flavor of their own" and there is not much kinship diversity generated by their mixed ancestry. We can see this by looking at the distribution of pairwise kinship in the Uyghur sample in Figure 4.5.

The U.S. African American population is, like the Uyghurs, a mixed population, but the mixture is only several centuries old. The left panel in Figure 4.4 shows that ancestral contributions to individuals vary wildly: some members of the population are nearly all African whereas at the other extreme some are

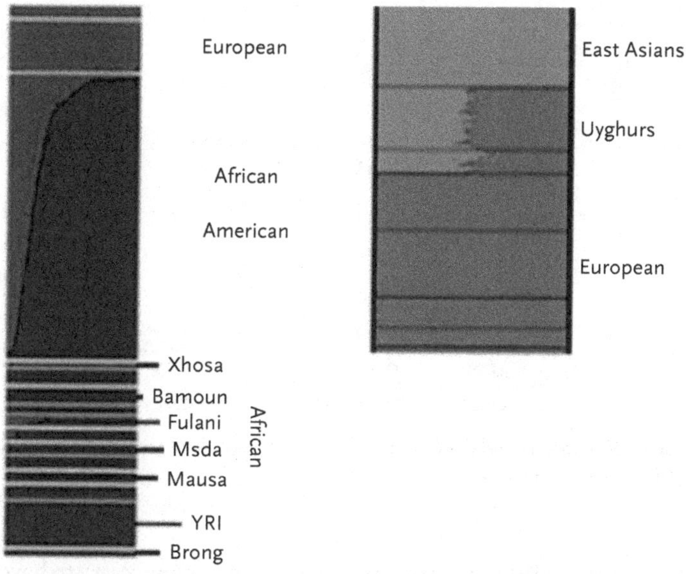

FIGURE 4.4 Ancestry components in two recently mixed populations. Each horizontal line shows the estimated ancestry of a single individual from each of the ancestral populations.

Source: http://blogs.discovermagazine.com/gnxp/2010/11/what-intra-inter-population-genetic-variance-tells-us/#.VthTLGuF9kl.

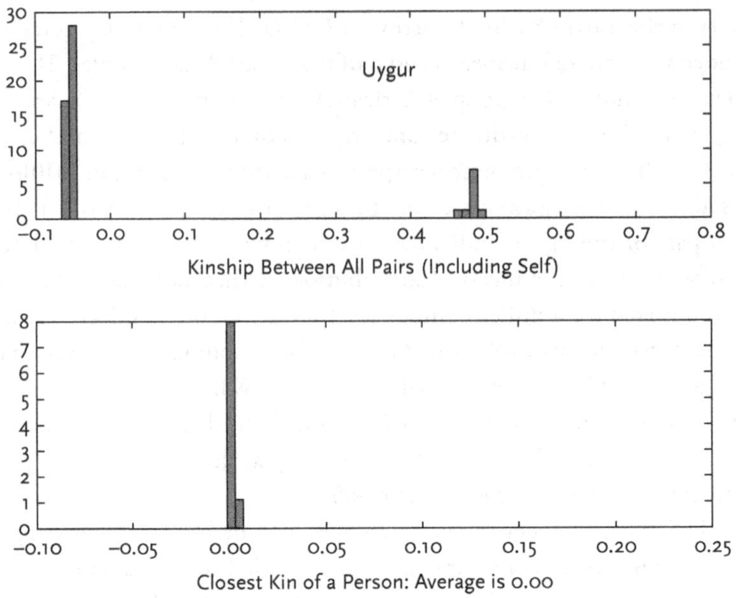

FIGURE 4.5 Kinship between all pairs of Uyghurs. See the caption for Figure 4.1 for an explanation.

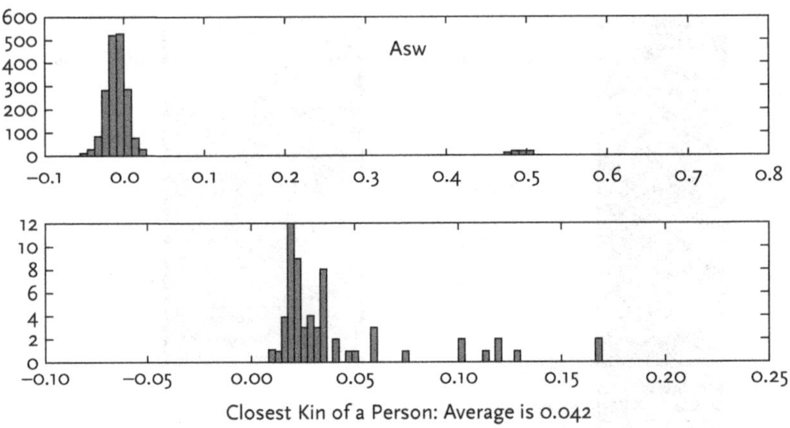

FIGURE 4.6 Kinship between all pairs of U.S. African Americans. See the caption for Figure 4.1 for an explanation.

nearly all European. There has not been enough time for ancestry to have become homogeneous within the population. This is apparent in the pattern of pairwise kinship shown in Figure 4.6.

In this case, the picture is very different from that of the Uhgyurs. The distribution of pairwise kinship is broad: kinship between individuals varies from

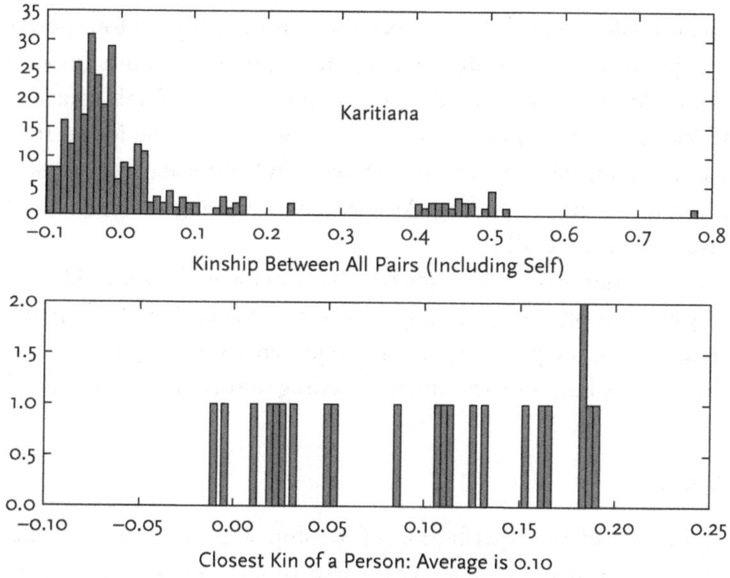

FIGURE 4.7 Kinship between all pairs of Karitiana. See the caption for Figure 4.1 for an explanation.

nearly grandchild to minus grandchild. The second panel shows, for each person, his or her closest kin as assessed from the whole genome. The average person can find someone related nearly as great-grandchild "worth," in Hamilton's calculus, one-eighth of himself. This population invites kin identification, kin favoritism, and nepotism. Here there would indeed be selection for kin recognition, probably first for apparent markers such as skin color.

A similar pattern of widely dispersed kinship with a group is seen in small, isolated populations that have little or no gene exchange with outside groups. Figure 4.7 shows pairwise kinship among a sample from the Karitiana, an isolate of 320 people in Brazil. In this case, recent admixture is not responsible for the pattern; instead, the shallow depth of genealogies in such a small group leads to the dispersal of kinship. As in the case of the recently admixed U.S. African American population, this is fodder for nepotism and discord within the group. Although there are no available data about Yanomamo villages, We expect they would show a pattern much like this one.

Assortative Mating

Humans in North America and Europe are known to mate assortatively. That is to say there is a weak positive correlation between mates for traits such as IQ, stature, skin color, and others. Some have speculated that this "like marries like" phenomenon reflects matching between mates for kinship. In four populations for which public data are available, there are typings of parents (mothers and fathers of children in the sample). These populations are Utah Mormons, the U.S. African American population from the Southwest, the Masai from East Africa, and a sample of Mexican descent living in California. In none of these is there a hint of assortative mating. As assessed by whole-genome assays, there is no assortative mating at all.

There is an argument to be made that selection would favor mild assortative mating by kinship, but we apparently cannot and do not do it. Perhaps observed assortative mating for phenotypic traits represents some vague inclination to match by kinship; but, if so, the effort according to this small sample is a failure.

Conclusion

This exploration of the distribution of kinship within populations has, in a sense, simply told us what we already knew. In large social groups there is little or no fitness payoff to kin identification, ethnic strife, and clannishness. Anthropologists have long spoken of small, genetically isolated societies as "kin based," and perhaps we have measured why that should be. In such societies the "kin" in "kin based" refers to social and genealogical kinship, which does not require genotyping. Ethnic diversity has the potential to lead to ethnic strife and other unfortunate outcomes, and we also know this very well (Putnam 2007).

We are left with the puzzle of why humans find ethnic discord to be congenial. Under the Bushman model of the human past there would have been no selective pressure to favor kin, and the Bushman model seems to be at the center of much theorizing about our EEA. Under the Yanomamo model, with its kaleidoscopic sequence of expansions, contractions, and extinctions, there could have indeed been selection favoring genetic similarity, but perhaps not very strong selection.

Our own favorite theory is that ethnic strife, rather the ability to learn ethnic strife easily, evolved in the context of agriculture, sedentism, long-distance trade, and urbanism. We know that the characteristic time for evolutionary change in humans is on the order of millennia, so the necessary time depth is certainly available (Cochran and Harpending 2010). Still, if persistent, large, diverse social groups are the context for such evolution, we ought to see strife among duck species and races. Such strife is certainly not apparent in the swamps that we frequent.

An important, relatively recent realization in the social sciences is that evolutionary change in humans occurs over centuries and millennia. This is uncomfortable for many of us because it implies that human differences are likely to be in the genome and these genomes may differ a lot in time and space. If the difference between Caesar's legions and the Italian army in World War II is, in part, a reflection of gene differences, then the discipline of history will have to expand and to change. If the fierce bravery of the Pashtun and the pacifism of Gandhi reflect, in part, evolutionary and genetic change—something as simple as widespread cousin marriage and consequent clan division among the former—then genetic history must be part of our understanding of history. Churchill (1898) bemoaned the imminent loss of the fierce Pashtun spirit in the face of rationalism and machine guns, but recent events have proven him terribly wrong.

References

Bulmer, Michael. 1994. *Theoretical evolutionary ecology*. Sunderland, MA: Sinauer Associates.

Chagnon, Napoleon A. 1968. *Yanomamö, the fierce people*. New York: Holt, Rinehart and Winston.

———. 1988. Life histories, blood revenge, and warfare in a tribal population. *Science* 239: 958–992.

Churchill, Winston S. 1898. *The story of the Malakand Field Force*. Report 2006. Rockville, MD: Wildside Press.

Cochran, Gregory, and Henry Harpending. 2010. *The 10,000 year explosion: How civilization accelerated human evolution*. New York: Basic Books.

Daly, Martin, and Margo Wilson. 1988. *Homicide*. New York: Aldine.

Davies, Gail, et al. 2011. Genome-wide association studies establish that human intelligence is highly heritable and polygenic. *Molecular Psychiatry* 16: 996–1005.

Hamilton, William D. 1964a. The genetical evolution of social behaviour: I. *Journal of Theoretical Biology* 7: 1–16.

———. 1964b. The genetical evolution of social behaviour: II. *Journal of Theoretical Biology* 7: 17–52.

———. 1975. Innate social aptitudes of man: An approach from evolutionary genetics. In Robin Fox, ed., *Biosocial anthropology*, 135–155. New York: Wiley.

Khan, Razib. November 8, 2010. What intra- & inter- population genetic variance tells us. http://www.unz.com/gnxp/what-intra-inter-population-genetic-variance-tells-us/.

Lee, Richard B., and Irven DeVore. 1976. *Kalahari hunter-gatherers: Studies of the !Kung San and their neighbors*. Cambridge, MA: Harvard University Press.

Li, Jun Z., et al. 2008. Worldwide human relationships inferred from genome-wide patterns of variation. *Science* 319: 1100–1104.

Milner, George R., Eve Anderson, and Virginia G. Smith. 1991. Warfare in late prehistoric west-central Illinois. *American Antiquity* 56: 581–603.

Nelson, Dean October 2012. India's dwindling Parsi population to be boosted with fertility clinics. *Telegraph*, World News, Asia, India. http://www.telegraph.co.uk/news/worldnews/asia/india/9612009/Indias-dwindling-Parsi-population-to-be-boosted-with-fertility-clinics.html

Putnam, Robert D. 2007. *E pluribus unum*: Diversity and community in the twenty-first century: The 2006 Johan Skytte Prize Lecture. *Scandinavian Political Studies* 30: 137–174.

Wright, Robert. October 2012. New evidence that racism isn't natural. *Atlantic.* http://www.theatlantic.com/health/archive/2012/10/new-evidence-that-racism-isnt-natural/263785/

Yang, Jian, et al. 2010. Common SNPs explain a large proportion of the heritability for human height. *Nature Genetics* 42: 565–569.

Yellen, John, and Henry Harpending. 1972. Hunter-gatherer populations and archaeological inference. *World Archaeology* 4: 244–253.

5 DARWINIAN EVOLUTION OF FREE WILL AND SPIRITUAL EXPERIENCE

Michael R. Rose

The Darwinian Problems of Free Will

The free will problem has as many different aspects as there are theories of human behavior, whether philosophical, ethical, religious, and so on. My perspective here is solely that of an evolutionary biologist. One main Darwinian problem with human free will is that our behavior is both extremely complex and not obviously related to Darwinian fitness. *Homo sapiens* exhibit truly distinctive elaborations of tool use—to such an extent that we are now virtually surrounded by artifacts of our own construction. Coupled with this rampant tool use is a degree of behavioral plasticity that is unique, evolutionarily. We have the capacity to learn how to accomplish an amazing variety of tasks and then to learn new tasks. No other organism on this planet comes close to our staggering potential for novel behavior.

Human beings make numerous choices and carry out complex behaviors in the course of their everyday lives, *apparently* unconstrained by genetic controls on most of their behavior. The tremendous diversity of both behavior and subjective experience, among individuals and cultures, further underscores an ostensible freedom that characterizes human behavior in contradistinction to the relative uniformity and genetically constrained behavior of all other animal species.

But how did evolution by natural selection produce a species that is, at least seemingly, no longer constrained by obvious Darwinian imperatives genetically encoded into our genomes? Extreme plasticity is dangerous from the perspective of evolutionary fitness. Wrong choices can take us far from fulfilling the Darwinian mission to reproduce. With our remarkable capacity to invent novel behaviors, what stops us from usually going awry?

The evolution of human free will has always been a puzzle for the Darwinian theory of evolution. An excellent argument could be made that natural selection should actively oppose the evolution of behavioral free agency. Thus, an exceptional evolutionary scenario is required to explain how free will evolved by natural selection. This chapter concerns just such a scenario.

One way to escape this problem is to invoke the rather bland notion that we have somehow "escaped" biological evolution. Thus, it is often supposed that a large, efficient, open-ended intelligence is something that we have been given by hominin evolution, and now we are free to use it as we please. This is an attractive theory. It immediately makes sense of the lack of Darwinian focus to most of our thoughts and emotions. By supposing from the outset this masterful stroke of being free of evolution, we thus have a free will that has nothing to do with Darwin, or the Darwinian reasoning in the publications of sociobiologists or evolutionary psychologists, particularly in the intellectual lumpenproletariat of popular science writing. It makes our free will something of existential power, possibly even beauty. It avoids any academic imperialism by the impingement of evolutionary biology on the traditional social sciences, as in the Promethean claims of the last chapter of Wilson's (1975) book *Sociobiology*, the one on human behavior.

This elegant solution to the problem of the evolution of human free will is unfortunately vulnerable to one devastating fact: selection against an upright mammal having a large brain is intense. In fact, such selection is one of the best documented of all forms of natural selection acting on *Homo sapiens* (Ellis 1973). This increase in brain tissue has been associated with many secondary transformations of human biology. The large head of the human neonate is a major cause of perinatal mortality, both for itself and for its mother (Ellis 1973). Or at least it was before the advent of routinely successful Cesarean sections allowed the survival of both mother and child. The human neonate is one of the most altricial of all mammals, requiring extensive care and protection to survive. After all, the newborn human is a highly dependent, immobile, vulnerable mammal. Furthermore, it remains that way for more than a year, during which it is usually not even able to walk. One interpretation of the human pattern of child development is that the human infant is really a fetus for its first 12 to 18 months. The evolutionary problem was probably the difficulty of giving birth to a one-year old human. Women's pelvises simply aren't big enough for that task. Even with the radical transformation of the early months of human development, the human female pelvis has been remodeled substantially in a manner that often interferes with locomotion at the limits of running speed to accommodate the large head of the neonate. The energetic costs of human pregnancy and lactation are considerable. Adult brain tissue is also the most costly tissue from the standpoint of both basal metabolism and volume of blood flow per unit weight; brain metabolism can use up to 40% of the calories expended by the human body, with an average around 20% to 25% (Leonard and Robertson 1993). The human brain "cost" a lot to evolve. It cannot be a neutral attribute.

The implications of this fact are fatal for the "escaped from natural selection" theory of human behavior. If our large brains are no longer relevant to fitness, yet are metabolically expensive and fairly deadly during birth, then they should evolve rapidly toward much smaller sizes, like those of chimpanzees. Our large brains must therefore have been maintained actively by strong natural selection favoring their use in ways that increase our Darwinian fitness, at least up until 1900, if not up until the present day. My conclusion thus is that the complexity and versatility of human behavior must have been sustained by powerful natural selection up until the 20th century. But, that pattern of natural selection clearly has been different from what has acted on almost all other species in the history of Earth. *Those* species have been selected for *very specific* behaviors that enhance their Darwinian fitness, much as sociobiologists suppose. We haven't. Specific behaviors have apparently not been the chief target of human selection. Our behavior seems to be somewhat free of Darwin's Procrustean evolution, *but we can't be*. The problem remains how our remarkably varied behavior evolved by natural selection, why we seem to have some degree or type of free will, and what that "free will" actually is. This is the first paradox that I propose to solve here.

The Mental Arms Race Amplifier

Our brains must have first evolved because of strong selection for some type of intelligence, regardless of whether present-day intelligence tests measure correctly the type(s) of general problem-solving capacity we were initially selected to possess. And from the argument presented earlier, this type of selection also must have been sustained up until the 20th century to forestall the rapid selective diminution of our brain size. (In my lab, we have made a point of showing experimentally how rapid and powerful reverse selection can be when there is antagonistic pleiotropy—the genetic manifestation of trade-offs [e.g., Teotónio and Rose 2000].) This must be the starting point for any significant treatment of the evolution of human behavior. Thus, an appropriate solution to the problem of the evolution of human free will must, at a minimum, take into account the selection mechanisms that produced us and have since sustained our behavioral flexibility.

Many theories of human selection for high levels of some type of intelligence involve specific scenarios for human ecology in Africa millions of years ago—bad weather, a reliance on hunting, the use of dexterous fingers to handle seeds, the behavioral self-control to sustain monogamy, a division of roles between the sexes, and so on (e.g., Lovejoy 1981). Although these theories are often ingenious, and may indeed correctly capture details of our evolution accurately, in the absence of time machines they are very difficult to test scientifically. Instead of pursuing such detailed and usually untestable scenarios, as I have done in the past (Rose 1998, Chapter 8), I lump past theories for the selection pressures driving human evolution into two broad categories: selection for "technical intelligence" and selection for "social intelligence." I also subsume within these categories specific behavioral adaptations: language, calculation of spatial location, managing one's personal database of friends or enemies, and so on. This is not to deny the existence of such specific functional components within the broad technical and social modes of brain function, where such specificity may involve both foci for selection and localization of such functions within the brain. For now, however, my intent is to bring forward salient empirical differences between the two main arenas for natural selection on human brain function.

Upright bipedalism and a fairly large brain gave early hominins, such as the australopithecines, the opportunity to learn how to use tools with their forelimbs, and our manual anatomy has evidently evolved as a result of this function. There is abundant evidence for the use of handheld tools over some millions of years of hominin evolution (e.g., McPherron et al. 2010). The use of simple learned technologies is now well-known among primates other than hominins, chimpanzees being particularly prolific tool users (Boesch and Boesch-Achermann 2000).

Discussion of technical intelligence in human behavior is of long standing—indeed, antedating Charles Darwin's theory of evolution by natural selection. The use of handheld tools to obtain food or to fabricate shelter is a dramatically distinctive feature of human life. Engels made the additional proposal that tool use for material purposes was key to human evolution in an unfinished essay of 1876 titled, "The Part Played by Labour in the Transition from Ape to Man" (1987). This idea has been advanced, burnished, and elaborated on many times since. In 1959, Oakley pointed out that tools made man "the most adaptable of all creatures," and that the use of tools may have been responsible for the evolution of human mental powers. Oakley did not offer a selection mechanism defining precisely how this occurred. In 1960, Sherry Washburn suggested that positive feedback for tool use led to consistent bipedality, resulting in a novel human ecology and consequently selection on many parts of the body that increased the ecological advantages of human tool use still further. A stone tool can be used for pounding, digging or scraping. A rock can turn a fist into a lethal weapon. Tools can also be used to make other implements, such as containers for carrying and storing food. Washburn thus proposed that the key to human evolution is our adoption of a tool-use ecological niche.

Other scientists, particularly anthropologists, have proposed theories of human evolution based on selection for some type of "social" calculation (e.g., Humphrey 1976; Byrne and Whiten 1988; Whiten and Byrne 1997; Humphrey 2003). In other words, it has been proposed that humans evolved substantial increases in intelligence primarily for the purpose of intraspecific conflict and cooperation. This is the social or Machiavellian intelligence model for human evolution. There are a variety of contexts in which social intelligence could be selectively favored: courtship; intrasexual competition for mating opportunities; competition for access to food, territory, and so forth. It takes little imagination to see this.

Extensive learned tool use provides a novel context for selection on social intelligence. Again, this idea has been broached repeatedly. Handheld tools that are used for hunting large animals can also be used to injure or kill other humans in armed combat. This is a central fact that conditions human evolution. A variety of authors have pointed out that armed combat could have generated intense selection for the intelligence to invent and use weapons with deadly force (e.g., Bigelow 1969). Moreover, armed conflict would create clear advantages to those individuals who adopt better tactics, including alliances for joint attack or group defense. This is a much better context for the invocation of Darwinian selection on social intelligence than that of gossip, to give a contrasting example, because armed combat can produce large effects on fitness, thanks to death, castration, and other misadventures of battle.

Consider the problem of two stags fighting for the opportunity to mate with a deer. In evolutionary game theory, this situation is analyzed in terms of alternative strategies such as Hawk, Dove, Bourgeois, and so on (Maynard Smith 1982). These are each candidate evolutionarily stable strategies (ESSs). ESSs arise from particular game contexts. In the case of a stag contest, this context includes how many deer there are to mate with, how sharp the antlers of the other stags are, how fragile antlers are, and so on. These are all biological variables that will, in turn, depend on the functional morphology, ecology, and physiology of the particular deer species. And these contest determinants cannot be altered deliberately by the contesting stags. Most animals can't break the rules, or "cheat," in their evolutionary games, because the game rules are set by the biology of their species. This is why evolutionary game theory works. Evolutionary games have stable rules because the overall contexts for conflict are stable within each animal species.

But, this constraint does not apply to hominins that use tools in armed combat. Among such hominins, the evolutionary game contexts in which they have their conflicts will not be determined stably by their general species biology. Unlike antlers, horns, claws, or fangs, hominin handheld weapons were not built-in; their design, fabrication, and use was not specified by genetic inheritance. As such, they would not establish a stable evolutionary game.

Thus, armed combat among hominins undermined the consistency with which their conflicts would produce an evolutionarily stable strategy. And the more proficient, complex, and culturally dependent armed combat within hominin species becomes, the less useful genetically fixed social strategies will be in such combat. [It should also be said that social strategies for armed combat would lead to selection for the capacity to take into account kin relationships and opportunities for reciprocal altruism. The current argument is not antagonistic to the view that cooperation and group benefits may have played a large role in human evolution.] Conventional evolutionary game theory becomes irrelevant to the prediction of the evolution of social behavior in such species, although this last problem is of more importance for academics than for evolving humans.

The problem for our evolving ancestors was that relative strategic advantage depended more on the facility with which novel and appropriate tactics could be improvised quickly in evolutionarily unstable situations, rather than on the evolution of a good standard strategy specifying the particular circumstances under which a standard set of tactics are to be used. In other words, selection on the human brain was not only about how well you could consistently use a spear, but also about how quickly you could use your strategic sophistication when deciding your next move given a complex set of variables, including opponents as or more sophisticated than you, as well as potential cooperators within your group in cases of intergroup conflict. Indeed, the formation of a nonkin alliance itself would be a singularly important context for the use of strategic sophistication. Although a panther goes through a process of deciding when and how to attack its prey, it is using a standard set of tactics; it does not have what some of us would call "out-of-the-box thinking."

Proficient use of deadly handheld weaponry was the explosive material that set off the evolution of high levels of social intelligence, I propose. On this theory, social stereotypy became a liability for our hominin ancestors, unlike the situation for birds, say, in which classic ESS behavior patterns are typical among conflicting individuals. The invention of new weapons or tactics in our ancestors thereby created a second-level evolutionary game. In a species that is being confronted continually with changes to the rules of its evolutionary games, the most successful game strategy is often one that is determined by direct calculation, not genetic specification. Armed combat selects for imminent calculation of social strategies—that is, it selects for versatile social intelligence. This is a reasonable first approximation to a theory for the evolution of free will.

By way of clarifying where we are in this argument, consider the following question: Why is greater flexibility arising from the learned use of handheld weaponry, including tactics, different from greater flexibility in negotiating dominance hierarchies and detecting or using deceit within a group? Why would weaponry introduce more instability than in-group social strategy selection by itself? The point of the mental arms race selection is that weaponry undermines the fixity of the parameters of the evolutionary games revolving around contested outcomes. Merely investing in the calculation of better social strategies would be selection within a parametrically stable social setting, in that the biology underlying the costs and benefits of alternative strategies remains fixed, in the absence of novel weaponry. Thus, colonial birds, which lead very intense social lives, have not evolved notable behavioral flexibility.

But before proceeding further with this attractive theory for the evolution of free will, there is a major problem facing mental arms race selection: the fitness cost of investment in strategy calculation. Having a very large brain is not free. And having a very large brain when the neonate passes through the pelvic opening of an upright biped is extremely hazardous. Therefore, opposed to any conjectural Darwinian benefits of open-ended strategy calculation are the Darwinian costs of an enlarged, metabolically active brain. Theories of social intelligence that ignore this problem are like fantasies about the benefits of space travel that do not incorporate the material difficulties and costs of rocketry.

In terms of evolutionary theory, investing in strategy calculation has the properties of frequency dependence that define evolutionary games. Investment in strategy calculation is a "bidding" or "display" game, in which the highest bid is expected to win whereas the lower bids lose. Together with John Haigh, I published an analysis of the expected outcomes from such calculator games and others like them (Haigh and Rose 1980). We called such bidding contests *evolutionary game auctions*. It was an important feature of our analysis that we did *not* assume there will be perfect detection of the bidding level of opponents. That is, we allowed the possibility of overinvestment in players with aggressive bidding strategies. This is the appropriate assumption when contest bids involve investments in the growth of material structures, such as brains, antlers, fighting limbs, and so forth, where such investments can involve biological materials, energetic resources, delayed maturation during a protracted period of growth, and so on.

The results of our analysis were somewhat surprising. If investment in open-ended strategy calculation is costly, in fitness terms, then the evolutionary outcome is a distribution of varied investment levels, with some players investing zero, but many others investing at a fairly high level, up to a rough cutoff where the frequency of individuals investing at higher levels plummets rapidly to zero. This mathematical result was quite robust, applying to a wide variety of cases, thanks to a fairly deep analysis that was not dependent on specific functional features of bidding. Our findings imply that a mental arms race might increase the upper end of the distribution of investment in social intelligence, but it will not increase the minimum of that distribution. If this were the appropriate model for human brain evolution, then we should have numerous individuals in human populations with the intelligence of small-brained hominins, an intelligence arguably not much greater than that of contemporary chimpanzee species. My interpretation of this finding is that no mental arms race, by itself, could have generated the spectacular increase in brain size and intelligence that led to present-day humans. Thus, despite the intuitive plausibility of the many different mental arms race scenarios that have been proposed during the past 140 years, the hypothesis is not viable on its own.

But what if, for example, a social calculation brain function involving enemy detection also gives those who possess it the ability to detect dangers posed by predators? In this scenario, the costs of this "social" enemy-detecting capacity are evolutionarily paid for by its benefits for evading dangerous predators, such as large felines. Or, as another example, what if the development of greater analytic capacity for the most deadly penetration with a weapon in our armed combat gave us a comparable ability to adapt our use of spears to penetrate the bodies of our prey? In either case, the mental arms race in our social calculations could be paid for by its benefits in other contexts. In the terminology of my 1980 article, these could be cases of "perfect amplification" of a mental arms race (Rose, 1980). The theoretical expectation, then, is that, for characteristics like a brain's generalized enemy-detecting capacity, the amplified mental arms race should produce rapid increases in enemy detection, which would then be manifest anatomically as a small increase in total brain size.

Imagine, now, a number of these general-purpose brain functions, useful for both ecological purposes and social competition. The list of these brain functions might look like this: enemy detection; tactical improvisation; sequential planning; attention to side effects of choices; empathic modeling of the minds of conspecifics, predators, and prey; and so on. The ecological benefits of such general-purpose calculations of fitness-relevant contingencies, and their effective use in the orchestration of behavior, could pay evolutionarily for their use in the mental arms race. Under these conditions, we can expect an explosive increase in generalized intelligence, with an associated increase in brain size during the evolutionary process.

The flaw in this model is obvious: it is unreasonable to expect that the ecological benefits of increased investment in open-ended calculation and the life-history costs that arise from producing a correspondingly larger brain would exactly balance over all levels of investment in intelligence. But this is where the features of the eventual evolutionary outcome of mental arms races come into play. Mental arms races, when there is a net cost to investment in such arms races, generate the smear of investment levels in social intelligence described earlier, starting at the point where the increased brain size and activity become costly in its collateral effects on Darwinian fitness.

Imagine, therefore, a fluctuating cost function for each particular component of social intelligence, in terms of its utility in nonsocial problem solving. That is, suppose that the cost function sometimes has a positive value and sometimes a negative value. When further increases in brain investment are of net benefit outside the social competition, the mental arms race will be accelerated by ecological selection. With a perfect balance, the mental arms race will be cost-free, and will proceed quickly. But when there is a range of brain investments that are costly in their net effect on fitness outside of the mental arms race arena, the mental arms race will generate a spreading out of brain investments from that point upward. If, at still higher levels of brain investment, the net effect of the brain investment outside of social competition becomes beneficial again, the mental arms race could "bridge" the range of investment values over which the investment is costly. Such "other-side-of-the-bridge" individuals will reap great rewards again, and the mental arms race will be accelerated once more, because these individuals will be smart enough to prosper both socially and ecologically, giving them much greater Darwinian fitness. In effect, they will be like Darwinian "overlords."

This is the mental arms race amplifier theory. It predicts the evolution of extremely high levels of investment in generalized brain functions—brain functions that must be useful for both social problem solving *and* nonsocial problem solving—for both courting a mate and tricking prey, for example. And it is very important to understand that *only* such generalized problem-solving capacities would be selectively favored by this bit of evolutionary machinery as brain sizes and metabolic levels increase to higher and higher levels. Hypertrophied but costly brain functions that enhance fitness only via social or technical intelligence, but not both, are not expected to evolve to high levels, although some such specialized brain functions could certainly evolve to a lesser extent. Thus this model predicts the evolution of a generalized intelligence, *not* one made up of hypertrophied problem-solving components that are specific to particular forms of social or technological problem solving, *not* focal forms of intelligence that are "locked on" to specific or stereotypical patterns.

Let me summarize where we are now, in my opinion. The key step for the evolution of human free will occurred when our ancestral hominins developed tools that destabilized their evolutionary games, such tools also being useful for acquiring food, defending against predators, and so on. Other species have weaponry that is "built-in," so their evolutionary game selection acts to generate well-defined behavioral strategies. But with handheld hunting implements, such as knives or spears, our ancestors lost stable evolutionary game rules. There was no way for natural selection to fashion specific, genetically encoded, behavioral strategies for our social behavior, because the most successful strategy then depended on the incidental cultural evolution of our use of weaponry. This generated an arms race for spontaneous calculation of social behavior relatively unspecified by genetics—indeed, forms of calculation that were progressively more innovative and unpredictable. Naturally enough, a capacity for creative calculation that could be used against conspecifics could also be used against potential prey, or indeed our former predators. This unusual nexus of circumstances produced the mental arms race amplifier pattern of selection, which in turn led to our evolution of a particular type of free will, regardless of whether philosophers consider such Darwinian free will relevant in any way to their metaphysical notion of Free Will.

Executive Multicamerality

A natural inference of the mental arms race amplifier selection mechanism for general-purpose, flexible Darwinian calculation is that humans rationally calculate the fitness contingencies of their actions and then choose that course of action that increases their Darwinian fitness. Many features of human behavior seem superficially to be reasonable corollaries of this theory, particularly the behavior of humans in market economies, where the common invocation of "self-interest" by economists sometimes seems like just one remove from a "fitness interest" as the source of human decision making. Likewise, the greater inclination toward promiscuity among human males compared with females is readily explicable in terms of rationally calculated Darwinian behavior, *if* indeed we do have rationally calculated Darwinian behavior. (I am not proposing here that this is the strongly held view of any present-day Darwinian; this is just a falsifiable "straw" theory.)

The Popperian beauty of this particular hypothesis is that it is eminently testable. But on any reasonable interpretation, this theory must be wrong. Much human behavior does not fit a rational Darwinian interpretation: vasectomies, substance abuse, homosexuals who are averse to heterosexual activity, vows of celibacy, and so on. But the most important empirical evidence against this theory is that people who aren't sociobiologists are not introspectively aware of their rational Darwinian calculations. Or if they are, then they are remarkably reluctant to disclose the fact.

In the afterword to his book *The Evolution of God,* Robert Wright (2009) has made this point quite directly:

> Even in our species—smart, as species go—the Darwinian logic isn't conscious logic; we don't go around thinking, "By loving my daughter I'll be more inclined to keep her alive and healthy until reproductive age, so through my love my genes will be playing a non-zero-game with the copies of them that reside in her." (pp. 457–458)

Even if some economists would rather humans were perfectly rational, or merely rational in a Darwinian, rather than strictly "economical," sense, we don't actually manage to be such a species, as the field of behavioral economics has shown frequently. That is, if we are supposed to be fundamentally rational in our decision making, then we must be remarkably poor at it, given the abundance of recent findings about the propensity of humans to make poor decisions.

Thus, a significant empirical problem is the lack of evidence that individual people adopt particular behaviors because of anticipated Darwinian consequences, leaving aside a few avid sociobiologists who might deliberately seek out opportunities for reproduction in conformity with their perceptions of Darwinian dictates and contingencies. Otherwise, we don't subjectively experience a process of Darwinian calculation within our mind.

But there is a remedy for this problem: executive multicamerality. That is, the existence of multiple centers of executive deliberation and decision making within the architecture of the brain's computational software. With such a system of orchestration of the brain software, multicameral species would have the advantage of being able to divide up different forms of calculation among distinct camerae. One center might make longer term strategic calculations whereas another handles the immediate decisions of implementation. Or one executive camera could be responsible for appetitive initiation of behavior whereas another is responsible for the inhibition of behavior.

The normal verbal disclosure of human subjective experience is that there is a single mind. I interpret this as indicating that the verbal center is disclosing the presence of one executive camera that is accessible to it. That our subjective experience does not disclose overt Darwinian calculation in turn suggests to me that one or more executive camerae other than the executive function that generates our consciousness discharges the crucial function of coordination of our behaviors to Darwinian ends. (This idea is not a new proposal; it is obviously akin to the Freudian concept of the superego.) Therefore, we must have at least two executive camerae, if this line of reasoning is accepted.

The general idea of multicamerality may be contrasted with the views of Antonio Damasio (1994) and many other neurobiologists who seem to be committed to the view that the only possible organization for human brain function is one executive consciousness and many subsidiary, simple, unconscious processes. If they remain committed to this view, then they have to deal with the evolutionary biological anomaly that our behavior is neither genetically focused on, nor consciously organized around, Darwinian goals. But the seeming anomaly of a lack of human, conscious Darwinian calculation processes can be resolved elegantly by the additional hypothesis that, in our species, Darwinian free will is embodied by a divided executive camerality, whereby the Darwinian free will is distributed among two or more major centers of executive calculation, with some of these camerae lacking access to the calculations of other camerae. (This is not to be confused with the subconscious, which serves as the substratum for the generation of the elements of human conscious decision making. All executive decision making is dependent on the aggregation of lower level information, in any hierarchical software design.) Although there is no logical necessity to such divided camerality among species with Darwinian free will, I propose that it has in fact evolved among hominins, perhaps as the result of a long-standing multicamerality in the much more focal behavioral programming of other mammals.

If the human brain operates multicamerally, with multiple higher order executive functions that have been sustained by evolution, who or what is the "self" that we subjectively experience? I propose the brain operations that produce our subjective selves constitute only one of *several* major suites of integrating brain functions—our multiple executive camerae. More specifically, I propose that our experienced self is specifically the immediate tactical coordination center for our behavior. To use a simile, our conscious minds are like the pilot on the bridge of a large ship. But the pilot is not in command. The pilot takes orders from the ship's captain. We are not, in fact, free to choose the meaning of our lives, if the "we" refers merely to the tactical center, or the pilot, which generates our subjective consciousness.

Spiritual Experience

How does our multicameral brain work to keep us entrained to Darwinian ends? I suggest that our subjective self is guided by means of sustained affect, directed perception, and long-term fixations. But this guidance of the subjective self does not follow any simple reflex or hydraulic pattern, contrary to Freudian theory.

Rather, our guidance toward Darwinian ends must be roughly as sophisticated as our subjective selves are. Otherwise, it couldn't keep our bumptious free will functioning in a manner that is oriented toward Darwinian goals.

Instead, in our species, the hypothesis proposed here is that overall guidance is supplied by an elaborate, unconscious executive center that in fact directs us as to our "life destinations," even if it leaves the specific navigation to the piloting consciousness. In other words, our conscious selves are the immediate operators of our bodies, but not the source of coherence, direction, motivation, and purpose in our lives. Our drives, pursuits, interests, and actions are generated by this elaborate executive center, although we are not directly aware of it. Thus we rarely make choices as "truly free" individuals. Instead, decisions regarding the direction of our lives are made for us by strategic centers that operate to supervise our conscious selves.

This doesn't mean there is another "person" inside our brain. But there is another kind of mind in our brain, one very different from the mind that we experience ourselves to be from moment to moment. The degree of unity that this "superconscious mind" possesses might well vary. (This terminological usage is parallel to that of Freud's "superego," with the qualifications indicated earlier.) However, in psychiatrically normal individuals who are not in a temporarily "altered state," whether of psychosis or intoxication—more on these cases later—my guess is that the strategic executive function has a degree of coherence. In some respects, it must have more coherence, persistence, and focus than our conscious selves possess. This other mind is the guarantor, the master controller, of our conscious self, keeping us entrained to Darwinian ends, despite our subjective experience of a wholly unconstrained will.

Given this hypothesis of a supervisory Darwinian brain function, there should be detectable side effects of its operation, even if it is separate from conscious function. Spiritual experience provides abundant evidence for such side effects. Religious movements and ideas generally involve the supposition and evocation of a secondary, or "spiritual," realm. This spiritual realm is furthermore often supposed to be populated with normally invisible entities that are involved somehow in human lives. Such suppositions may involve the literal invisibility of entities that lurk about our immediate environs, or they may suppose some type of invisibility arising from absence, because of the placement of these entities in another location (such as the top of Mount Olympus, Valhalla, or the Biblical heaven) or at another time (during the origin of the world or "ancient times," for example). Although the diversity of spiritual experience precludes any simple characterization, such experience clearly points to aspects of human existence that are not usually straightforward or immediately accessible to consciousness, yet somehow concern the moral appraisal or ethical significance of our behavior.

Indeed, a species that lacked spiritual experience entirely would be a doubtful candidate for multicameral Darwinian free will.

Additional evidence for the action of Darwinian supervisory functions in generating spiritual experience comes from cases when human brain function has been impaired by means of genetics, pharmacology, or extreme physiological stress. Under such conditions, many individuals experience contact with other "person(s)" or "will(s)" of some type. The natural inference, on the present theory, is that these are cases in which isolation of the executive camera has broken down, giving rise to direct conscious experience of our multicamerality.

Among the prominent features of psychosis are religious hallucinations and delusions. Delusions of being specially "chosen" and hallucinations in which deities appear are common features of cinematic and fictional renderings of psychosis, and they are prevalent among medical case reports concerning psychiatric patients. Similarly, in altered states of consciousness induced by psychedelic drugs or extreme physical stress, reports of "seeing God" or hearing "the voice of God" are common.

Why are such experiences so frequent when the normal limits of cognition and emotion are transgressed? My interpretation is that such experiences reflect a breakdown in the blockade that normally forestalls the conscious experience of Darwinian supervision by the superconscious mind. In other words, hypertrophied spiritual experience during delirium, intoxication, and psychosis is a more overt, although less functional, manifestation of our endogenous strategic controller(s) that are the source of all genuine spiritual experience.

Note that, for the tactical and strategic executive camerae to sustain distinct functions, they must have a significant degree of separation in their information processing. If tactical and strategic information processing were fully integrated, then the species would be unicameral. Thus, there has to be some type of separation between these two executive functions, leading to a lack of full accessibility of the contents of each function to the other, when there is multicamerality.

Spiritual experience, as documented historically by William James and as is now known by many people who have experimented with various religious denominations and cognate "New Age" practices, is much more than mere religious observances. That is, spiritual experience stretches well beyond formalized worship as part of a congregation. Indeed, it is important to note that formal religious observances can, in fact, be a sham, at least for some of the participants. Thus, the more pertinent type of spiritual experience is tantamount to some type of experience of a god, spirit, or merely an obscure "presence." Certainly this type of experience is not universal, any more than the appreciation of art or science is universal. But it is quite common.

A striking feature of intense spiritual experiences across cultures is the extent to which they have common elements. There is typically a sense of "presence." That is, there is a sense of another person, or somehow a personalized "force" of some kind, communing with the person having the experience. This is obviously the case with the type of prayer integral to Christian and Muslim experience. One is supposed to establish some type of "contact" with God in those faiths, contact in which one's wishes and hopes are expressed. Often prayer is seen as a means of obtaining wisdom. In such cases, the praying person appeals to "The Lord" for a specific type of understanding. Thus, a person might pray for an understanding of a difficult person, or situation, in his or her life, in an interesting anticipation of psychotherapy.

In mystical traditions, efforts to experience God or gods are still more extreme. Sufis whirl to achieve a mystical state in which they believe they commune directly with Allah. A variety of New World, non-Biblical spiritual traditions use hallucinogenic drugs, such as peyote, to achieve a state in which people commune with gods or the spirit world generally. Voodoo practice seems to feature an elaborate ritual of hypnosis as a prelude to direct contact with the spirit world. Even within the cultures that share a Biblical heritage, there are a variety of spiritualist rituals that are not usually encouraged by the religious authorities, such as séances, which are experienced as avenues of contact with another realm.

From the standpoint of multicameral Darwinian free will, a give-away feature of organized religions is that they characteristically offer strong claims or prescriptions concerning (a) prospects for immortality and (b) appropriate sexual conduct. And many religions and related ethical systems characteristically feature a component of what is broadly referred to as (c) "ancestor worship." All three of these things have strong Darwinian significance. Ancestor worship has an obvious Darwinian salience, in that it bears metaphorically on the fact of genetic variants propagating themselves from generation to generation. That is, suppose that religion is indeed a cultural phenomenon derived from the difficult relationship between a tactical consciousness that follows the dictates, directives, and desires instilled in it by a strategic brain function, which itself in turn makes the ultimate Darwinian computations. With this supposition, the religious obsessions with eternal life and sexual fulfillment, whether the chaste sexuality of the Christian creed or the orgiastic rites of some variants of Taoism and Hinduism, are only to be expected.

It might, therefore, be useful to distinguish clearly the present explanatory theory of religion from that of Tiger and McGuire (2010). For them, "God is a creation of the brain" (p. 215), although my view is that the spiritual experience of god(s) is a *feature* of the brain, a predictable result of our brain's software. Two further quotations suggest, in my opinion, considerable difficulties with their thesis. As an anthropologist and as a primatologist they admit, "Chimpanzees seem to display no acknowledgment of an identifiable and potent deity/being that directly affects their behavior" (p. 206), yet humans, who presumably would have a much greater ability than chimpanzees to dismiss a crude psychological ploy, characteristically suppose there "is an unseen choreographer in whose ballet they are happy to dance" (p. 194). Is it not the more elegant hypothesis to suppose there is indeed such "an unseen choreographer?"

Acknowledgments

I am grateful to the organizers of the Consilience meeting for the opportunity to voice my heretical views about human evolution. I am also grateful to Blanca G. Cervantes, Joseph C. Carroll, and Henry Harpending for close readings and critical analyses of early drafts.

References

Bigelow, Robert S. 1969. *The dawn warriors: Man's evolution toward peace.* Boston: Little Brown.

Boesch, Christophe, and Hedwige Boesch-Achermann. 2000. *The chimpanzees of the Taï Forest: Behavioral ecology and evolution.* New York: Oxford University Press.

Byrne, Richard, and Andrew Whiten, eds. 1988. *Machiavellian intelligence: Social expertise and the evolution of intellect in monkeys, apes, and humans.* Oxford: Oxford University Press.

Damasio, Antonio 1994. *Descartes' error: Emotion, reason, and the human brain.* New York: Putnam.

Ellis, Walter S. 1973. Mortality and birth weight in Philadelphia blacks: An example of stabilizing selection. *American Journal of Physical Anthropology* 38: 145–149.

Engels, Friedrich. 1987. The part played by labour in the transition from ape to man. In A. K. Avelichev et al., eds., *Anti-Dühring, Dialectics of Nature,* vol. 25 in *Karl Marx, Friedrich Engels, Collected works,* 452–464. Moscow: Progress Publishers.

Haigh, John, and Michael R. Rose. 1980. Evolutionary game auctions. *Journal of Theoretical Biology* 85: 381–397.

Humphrey, Nicholas K. 1976. The function of intellect. In Patrick P. G. Bateson and Robert A. Hinde, eds., *Growing points in ethology,* 303–317. Cambridge: Cambridge University Press.

———. 2003. *The inner eye: Social intelligence in evolution.* New York: Oxford University Press.

Leonard, William R., and Marcia L. Robertson. 1993. Evolutionary perspective on human nutrition: The influence of brain and body size on diet and metabolism. *American Journal of Human Biology* 6: 77–88.

Lovejoy, C. Owen 1981. The origin of man. *Science* 211: 341–350.

Maynard Smith, John. 1982. *Evolution and the theory of games*. Cambridge: Cambridge University Press.

McPherron, Shannon P., Zeresenay Alemseged, Curtis W. Marean, Jonathan G. Wynn, Denné Reed, Denis Geraads, René Bobe, and Hamdallah A. Bearat. 2010. Evidence for stone-tool assisted consumption of animal tissues before 3.39 million years ago at Dikika, Ethiopia. *Nature* 466: 857–860.

Oakley, Kenneth 1959. *Man the tool-maker*. Chicago: University of Chicago Press.

Rose, Michael R. 1980. The mental arms race amplifier. *Human Ecology* 8: 285–293.

———. 1998. *Darwin's spectre: Evolutionary biology in the modern world*. Princeton: Princeton University Press.

Teotónio, Henrique, and Michael R. Rose. 2000. Variation in the reversibility of evolution. *Nature* 408: 463–466.

Tiger, Lionel, and Michael McGuire. 2010. *God's brain*. Amherst, NY: Prometheus Books.

Washburn, Sherwood L. 1960. Tools and human evolution. *Scientific American* 203: 62–75.

Whiten, Andrew, and Richard Byrne, eds. 1997. *Machiavellian intelligence II: Extensions and elaborations* Cambridge: Cambridge University Press.

Wilson, Edward O. 1975. *Sociobiology: The new synthesis*. Cambridge, MA: Harvard University Press.

Wright, Robert 2009. *Evolution of God*. New York: Little, Brown.

III ANCIENT MARKINGS

6 NEANDERTHAL HUMANITIES

John Hawks

I study Neanderthals—ancient people who lived more than 30,000 years ago in Europe and western Asia. The dead speak to me only through their leavings. I have no Neanderthal informants. No matter how often I ask them questions, they will never answer. They left no video, no books, no inscriptions. They left stones and cut-marked bones, a few postholes and pits, bodies, and ash.

I struggle to gain insight into the humanity of these ancient people. Unquestionably they had rich social lives. No social primate can function without emotion, without affiliations or conflicts, without long memories of experiences with other individuals. No primate that depended on hunting and gathering, as all members of our genus have done, could survive without deep social bonds that enabled cooperation and mutual interest. Yet the Neanderthals remain enigmatic.

In a few exceptional cases, they carved regular lines upon objects. An optimist can point to a handful of clear indications of intentional behavior, giving the humanist some hope of treating them like texts that can, in principle, be interpreted. A pessimist replies that whatever "texts" remain are written in an undecipherable script.

What would Neanderthal humanities be like? It may appear that all attempts to build knowledge about the subjective lives of ancient people are necessarily fictional. But the question goes beyond the Neanderthals. Can we understand subjective experiences within other cultures? If so, what methods will bring us toward such knowledge of ancient people?

Throughout this chapter, I consider the problem of Neanderthal intentional markings. This is a rich area of inquiry in archaeology, with more and more evidence of Neanderthal "symbolic" behavior emerging every year. As much as we might wish it, intentional markings do not provide "windows" into the minds of ancient people. As described here, some past efforts to explain such markings cannot be distinguished from fiction. Yet, if the subjective lives of ancient people must forever remain unknown to us, then surely our attempts to interpret the lives of our contemporaries must be equally fictional. Consilience, as a general attitude about the ability of scientific knowledge to cross into humanistic realms, holds out the promise that we might study and come to an understanding of the humanity of ancient people, not only as social primates but also as subjects of experience. But I have little confidence that science can provide such knowledge, or that nonscience interpretive approaches can create knowledge distinguishable from fiction.

Anthropology is a diverse field. Biological anthropologists, like me, study the human organism and its evolution. Like all organisms, humans exhibit behaviors that may adapt them to their environments. For humans, the social environment is a primary determinant of the adaptive constraints shaping behavior. Our social lives are a niche that we construct continually; our conscious selves are tools for our nervous system to navigate this social environment. Subjective experience, in this way of thinking, is fundamental to human biology. Without it, we could not function in our social environments.

Many anthropologists study human behavior not from a scientific perspective, but within an interpretive frame. The interpretive anthropologist recognizes that a person's behaviors may not be fully translatable into different cultural contexts. Indeed, interpretation may be unsatisfying even when limited to the subject's own cultural context. Identifying the proximate or ultimate causes of behavior or belief is not the goal of interpretive theory; the interpretive anthropologist resists such reduction of subjective experience to physical causes. Theory, to the interpretive anthropologist, is a productive and creative activity, a synthesis of observation with the interpretive frame. By observing the subjective experience of another person and her own encounter with the literature, the interpretive anthropologist builds something new.

I consider myself a scientist. Some interpretive anthropologists reject scientific approaches and methods, describing them, at best, as irrelevant to understanding subjective experience—at worst, weapons of hegemony.

I consider myself a humanist. In academic discourse within anthropology, the interpretive approach is aligned with humanistic inquiry. My work is starkly different in its goals and methods from the interpretive anthropologist. Like many of my colleagues I pick my way through this minefield. Subjective human experience must have an evolutionary origin, which we seek to understand and discover.

Obviously, the broad concept of "consilience" is relevant to any interdisciplinary inquiry. Whewell's "*Consilience of Inductions*" is an important mode of argument in biological anthropology, just as in evolutionary biology (Whewell 1840; Gould 2002). Whewell (1858) held that consilience of inductions from different classes of facts "are the test of truth" (p. 2: 180). Edward O. Wilson's (1998) argument for consilience of different fields of knowledge is rather grander in scope than Whewell's conception:

> Consilience proved to be the light and way of the natural sciences. Physics, with its astonishing congruity to mathematics, came to undergird chemistry, which in turn proved foundational for biology. The successful union was not just a broad theoretical consistency, as articulated by Whewell, but an exact unfolding of principles pertaining to more complex and particular systems into the principles for simpler and more general systems. (p. 133)

Wilson discarded Whewell's reference to induction and argued that the fact of consilience is a logical consequence of the reduction of chemical and biological theories to physical theories. Consilience was itself, in Wilson's description (1998), substantially easier to establish than reduction. A scientist can observe that two statements at different levels of physical complexity are consilient without being able immediately to reduce the complex system to a simpler one. For example, the ratio of stable carbon isotopes in a mammal's tooth enamel reflects the photosynthetic pathway of plants eaten by the individual during the period of enamel formation. Eaters of grasses with a four-carbon photosynthetic pathway have enamel with a higher proportion of carbon-13 than eaters of fruit and leaves from trees with a three-carbon photosynthetic pathway. This hypothesis relies on consilience of theories underlying the physical properties of carbon isotopes, the chemistry of photosynthesis in plants, and the biology of enamel formation in mammals. Yet we cannot currently reduce the embryonic development of ameloblasts (enamel-forming cells) to physical principles. Nor can we synthesize, or build up, the actions of ameloblasts from chemical or physical theories. In practice, we make a much weaker assumption: these cells will not *betray* well-known principles or act according to their own idiosyncratic physics. Discussing reduction and synthesis, Wilson (1998) admitted the difficulty (and in some cases, impossibility) of building predictive hypotheses about complex systems from simple physical principles. He referred explicitly to the "paradox of emergence," which necessitates that levels of organization have their own proper explanatory theories.

Our attempts to understand retrospectively the results of complex interactions may demonstrate the fact of consilience among explanatory theories that apply to different levels of interaction. Does the fact of consilience prove that the explanatory theories at different levels reduce to a common set of simpler (presumably physical) principles? To the extent that we believe in a single material reality, we must conclude Wilson is correct. When we have theories at different levels that are more-or-less true, they should give rise to consilient inductions. If theories at different levels give rise to different predictions about the same physical event, they are, by definition, incompatible. If theories always give rise to the same prediction about every possible event, they are, by definition, identical. Consilience is no proof of reducibility, but it may serve as a sign of reducibility.

But what if our theories are not close to the truth?

The question may seem impish. After all, if our theories are so wrong, why would we bother? But the idea that similarity of predictions is an indication of similarity of theories, which seems plausible for the recent history of physical sciences, simply doesn't work in the social sciences and humanities. Replacing Newton's theory of gravitation with Einstein's was a tiny change for most easily observed physical systems, so tiny that we still teach Newton's theory to our children as correct. The history of the social sciences is a march through dozens of all-encompassing social theories with barely any attempt to falsify or test for their mutual incompatibility. Today's interpretive anthropologists flock among a confusing array of social theorists, each with different, partly incommensurable schemes of social explanation.

Consider, for example, the phenomenon of religious ritual. Counting only the past 200 years, theories of the social and psychological causes of religious ritual have been promulgated by dozens of major scholars, including Tylor, Spencer, Freud, William James, Durkheim, Boas, Campbell, and our contemporaries Pascal Boyer and Scott Atran. Most among this group have attempted comparative procedures similar in form to the consilience of inductions. The subject of religion almost demands such an approach, because it touches on so many aspects of human interactions and is understood through different religious traditions in different cultural groups.

An example close to the Neanderthals is provided by George Barton (1940), who argued that religious ritual remains as a legacy of prehistoric peoples' awe of the divine feminine:

> Palæolithic religion was, then, sex-mysticism. The psychologic unity of the race made it universal as its survivals in the historic period prove. This is the real origin of religion. It was not begotten by fear (Lucretius), nor by

> animism (Tylor), nor by ancestor worship (Herbert Spencer), nor by the mysterium tremendum (Otto), but by the mysterium feminium—a mysterium tremendum indeed, but scarcely that which Otto contemplated. (p. 131)

Barton (1940) turned to the study of myth and religion in small-scale societies, as well as the mortuary practices of Paleolithic people, to make the argument that ritual had its origin in the worship of females:

> Dr. Lewis has pointed out that in the case of such burials the body of the dead is covered with red earth and often either rests upon or is partly covered with the shells of various kinds of shellfish. He rightly infers that the image represented the mother-goddess—the goddess of life—that the red earth represented blood, the vehicle of life, and that the shells were emblems of the mother goddess because, like her, they contained life or produced a living creature. (p. 135)

Neanderthals never, to our knowledge, painted figures or other representations on the walls of caves. Nor did they accompany burials with masses of pigment as did some later European peoples. But some red marks on the walls of caves are now known to be contemporary with late Neanderthals, raising the possibility that they were the authors of this behavior. Pigment crayons, blocks of mineral pigment that have been worn by rubbing on some surface, have long been known from Neanderthal sites. A recent forensic analysis of such crayons has shown that some of them were rubbed on a soft surface, such as skin or hide (Soressi and D'Errico 2007). At two sites, small cup-shaped pieces of stony stalagmite have been found with traces of mineral pigment inside them. The Neanderthals appear to have been painting, or mixing mineral pigment for some other purpose. Evidence of shellfish exploitation is rarely found at Neanderthal sites, possibly because the lower sea level during Neanderthal times meant that shorelines were far from today's sites in most of Europe. Nevertheless, a handful of sites show that Neanderthals were using shellfish and other marine resources (Barton 2000), and at one site Neanderthals were gathering empty shells with natural holes, apparently stringing them together. One Iberian site preserves an exceptional shell bearing a natural band of pigment on one side and on the other, painted by a Neanderthal, a red ochre band mirroring the opposite side (Zilhão et al. 2010). With this kind of evidence emerging from the archaeological record of Neanderthals, perhaps we can test such hypotheses as Barton's (1940) for the origin of religious ritual. Barton's work does not stand alone; several scholars have pursued similar models for the origin of religion, relating red pigment, female symbolism, and the origin of religious ritual or broader social interactions (e.g., Wreschner 1980; Knight 1995). Menstruation in this view ties blood with the social power to govern sexual relations; women in a number of societies undergo an involuntary seclusion at the time of menstruation. Fertility is a mysterious and powerful process within many societies, and menstruation visibly connects fertility with blood. A connection of religious ritual and the sacred feminine reads like the *Da Vinci Code* of Paleolithic archaeology, depending on observations drawn from ethnography, folklore, political economy, art, and Paleolithic archaeology. In Knight's (1995) model, as an example, the appearance of complex social interactions, symbolic behavior, and language is explained by a sexual revolution in which women gained meat and protection from men by controlling sexual access. Knight and others have tied these phenomena to ethnographic and economic observations among African and Australian small-scale societies, tying them further to the iconography of rock art.

What should we make of this seemingly plausible hypothesis of a causal connection of religious ritual, menstruation, motherhood, shells, and the color red? I suggest the idea is an "umbrella hypothesis"—an attempt to explain a large set of true or uncontroversial propositions as results of a single underlying cause (Langdon 1997). An umbrella hypothesis is not necessarily false, or even problematic. But the strength of support of the hypothesis is not a function of the facts adduced in its favor. Indeed, the hypothesis does not *necessitate* that any of these propositions be true; if one was false, some *other* proposition might otherwise have been found as support for the hypothesis. An umbrella hypothesis subtends an internally consistent model of causal relations among facts.

How can we explain the association of red pigment with burials and Paleolithic art without a causal relation among menstruation, religious ritual, and art? Red ochre, the second most common pigment in archaeological sites after black, is much easier to obtain and survives better in archaeological contexts than most other color pigments. Likewise, shells are durable archaeologically, and collected and traded widely in nearly all preindustrial societies. Symbolic association of the color red with blood is ubiquitous, even inevitable. But other correlates of red vary widely across cultures: in some, generally positive; others, negative; some masculine and others feminine (Wreschner 1980). To be sure, some traditional African groups have placed enormous social and economic importance on red cosmetics, associating them with menstruation and status. These elaborate traditions leave heavy material traces from the acquisition, distribution, and use of red pigments, and not a trace of such elaborate material traditions can be found in the Paleolithic record. Instead, we have traces of red ochre use, scattered across much of the Old World during the past 300,000 years, with a handful of co-occurrences with shells.

We are not looking at a *consilience* of red ochre burials, shells, and mother-goddess ideation, we are looking at a *coincidence* of these things in some material and cultural contexts.

From a purely interpretive perspective, it may not matter that the coincidence is accidental. The strong association of menstruation and red pigment in the ideological systems of some societies is a fact that has braced the social lives of thousands of people across hundreds of years. People's subjective experiences within such cultural contexts have included these social facts by which people have navigated their social life. Interpretation of these facts is a form of play, sometimes reversing hypotheses of cause, connecting them to the work of social theorists, reading them as if they were a text through an interpretive frame. But to apply this frame outside the context where it is observed—to extend it to Neanderthals or other Paleolithic peoples—is fiction.

The scientist wants more than to interpret the associations where they occur; she wants to predict, to deduce, to infer unknowns. And for this, she needs far more than the observation of their occurrence together. She must show the associated facts to be causally connected. A statistical association between the elements of the hypothesis, the red pigment, shells, symbols of the feminine, and rituals related to menstruation in recent human cultures might test the hypothesis of causal connection among them. A consistent record of ideational connection from informant interviews might also contribute to such a test, or a deeper record of archaeological shell use by Paleolithic people. Such evidence helps to substantiate that the inference *as applied to ancient humans* may be more than fiction. Such evidence drawn from several different lines of inquiry might contribute to a consilience of inductions. Because such an approach must be tied to *more lines of evidence*, we must necessarily proceed *more slowly* than the purely interpretive anthropologist. We must also spend more time on each line of evidence, establishing the direction and strength of causal relations among observations.

More than 20 years ago, most archaeologists held that only modern human groups show clear evidence of intentionally marking objects for the purpose of communicating. The upper Paleolithic record of artistic expression is vast and unparalleled by anything that came before it. When a visitor today walks through a cave with art painted on the walls, she experiences a reality of iconic expression that simply did not exist anywhere in the world before 50,000 years ago. Painted art, sculpture, musical instruments, and other nonvocal modes of communication appeared in the archaeological record of Europe, Australia, and southern Africa 80,000 years to 30,000 years ago. Some archaeologists have dubbed this phenomenon "the human revolution," suggesting that such communication provides the first evidence that human minds had become truly subjects of experience (Mellars and Stringer 1989).

This view has become increasingly problematic as evidence of the capabilities of Neandertals and other archaic humans has grown (D'Errico 2003). As noted earlier, the exceptional evidence of intentional marking by Neanderthals contributes to this record. More evidence comes from the growing record of behavioral sophistication among African peoples of the Middle Stone Age period, earlier than 70,000 years ago. At several sites, including some relatively far from the seashore, pierced shells have been found (Bouzouggar et al. 2007). Ostrich eggshells were incised with geometric patterns in southern Africa as early as 80,000 years ago (Texier et al. 2010). A tradition of pigment use, including the complex preparation of mixtures, is of equivalent age in southern Africa as in Europe, in both cases dating long before the origin of modern human populations.

Faced with these facts, archaeologists have problematized the origin of marking behavior. At one extreme, some suggest we should maintain the null hypothesis that archaic people had cognitive capabilities identical to those of living humans. From this point of view, evidence of marking, pigment use, or the accumulation of nonutilitarian objects such as shells support the interpretation that the symbolic communication abilities of humans extend far back into the Pleistocene. At the opposite end of the scale, we can observe that many of the archaeological cultures of the past 10,000 years fail to show systematic evidence of marking, gathering of exotic objects, or other archaeological indicators of nonvocal communication (Speth 2004). The lack of such evidence does not indicate that these recent peoples were incapable of such activities; it merely reflects an ecological or social context in which people did not value the production of such material objects.

Archaeologists have begun to consider models for the gradual emergence of such cognitive abilities in archaic humans. Within Africa, the sophistication and standardization of stone tool technology increased across the past 300,000 years. The appearance of small projectile points, long-distance movement of raw material, and increasing pigment use during the past 100,000 years seems a logical extension of a much longer term process (McBrearty and Brooks 2000). Likewise, European Neanderthal populations show a gradual increase in sophistication of tool raw material acquisition, stronger evidence of a diversity of material culture, and pigment used in marking behavior during the same time period. Just as genetic evidence has made the genealogical connection between living people and archaic people more complex, so the archaeological record has made the interpretation of behavioral capabilities in these ancient people more complex.

Our growing knowledge of Neanderthal marking behavior shares little with the interpretive approaches of cultural anthropology. We see only hints and traces of the behavior of a few Neanderthals at a handful of instants in their lives. With the cave art of the Upper Paleolithic, the art historian can use interpretive methods to add value to our understanding. She may not be able to read the minds of the artists, but she can arrive at an understanding of their stylistic processes, the relation of representation to their lives, and the picture of their social existence. Archaeologist Dale Guthrie (2006) brings a contextual understanding to Paleolithic art built on his experience as a hunter and his consideration of Paleolithic groups. The consistent representation of what he calls "high-testosterone activities" helps to place the role of the artist within their hunting society. The representations of animals attend to small details of their biology—such as rutting male deer, animals at the peak fat distribution of autumn, and prowling carnivores—testify to the intimate ecological knowledge of the artists. Meanwhile, the distribution of handprint sizes shows that children were often among the artists, an inference supported by the relatively crude form of a large majority of representations seldom illustrated in books about cave art. The understanding attained by the art historian about these ancient people is not dialogic, not negotiated, nor is it based directly on testimony. But it is undeniably humanistic.

With Neanderthals and their contemporaries we can approach their humanity only through a fog. Drips of red ocher in the dust of an ancient site do not speak of their maker (Roebroeks et al. 2012). They do not even have a maker in the sense of an author of intention behind their production. Repeated marks on a bone may reflect intention, but ascribing intention requires ruling out alternatives that may, in practical terms, be impossible to exclude for a singular object.

Still, our inability to arrive at a symbolic or iconic understanding of Neanderthal production does not preclude a truly humanistic understanding of them. A humanistic interpretation of the cave art of the Upper Paleolithic cave art depends on much more than the content of the images themselves. Such an understanding can be built by juxtaposing the content of the art with a sophisticated knowledge of hunting practice tied to a depth of ethnographic knowledge about human foraging and social interactions in small groups. When we have a blind spot of interpretation, it is because we assume too much, not because the paintings tell us too little. In other words, humanistic understanding in this context does depend on a consilience of inductions in the Whewellian sense.

The remains of the Neanderthals and other ancient peoples do speak to us. We can develop some knowledge about the subjective lives of these people, but this knowledge is necessarily limited and can be attained only very slowly.

Anthropologists approach the problem of understanding subjective experience in two ways. Scientific anthropologists approach the problem with great caution, proceeding slowly and by degrees, building understanding of small parts of the overall picture. Interpretive anthropologists exercise implicitly even greater caution by assuming that subjective experience may never be comprehensible even in principle. They therefore focus on the aesthetics of an improvised encounter between observation and theory.

I do not deny that the interpretive approach can yield interesting and compelling stories. The methods and style of humanistic and interpretive anthropology are not sterile; they actively bring new perspectives on the observations made by ethnographers and other cultural anthropologists. But it is not in the nature of such interpretive methods to comport with the requirements of hypothesis testing. I assert that these approaches really cannot be reconciled, in the Wilsonian sense of consilience. The aims of interpretive anthropology are productive and aesthetic in nature. This is not a misunderstanding within anthropology, it is a real difference of priorities. Interpretive anthropology cannot be reduced, even in principle, to scientific principles of biology or psychology.

Sometimes I despair that the noise of exuberant interpretive work may drown out the quiet signal as we slowly build up multiple lines of evidence about the lives of ancient people. We are building a humanistic understanding of Neanderthals and other ancient people. We may never know the songs of Neanderthals, but we will someday know whether any songs could have existed. We may never find a Neanderthal portrait, but we will assemble an understanding of how pigment marking once fit within now-extinct social systems. It diminishes the humanities to restrict them to mere interpretation.

References

Barton, George A. 1940. The Palaeolithic beginnings of religion: An interpretation. *Proceedings of the American Philosophical Society* 82: 131–149.

Barton, Nicholas. 2000. Mousterian hearths and shellfish: Late Neanderthal activities on Gibraltar. In Chris B. Stringer, Nicholas Barton, and Clive Finlayson, eds., *Neanderthals on the edge: Papers from a conference marking the 150th anniversary of the Forbes' Quarry Discovery, Gibraltar*, 211–220. Oxford: Oxbow.

Bouzouggar, Abdeljalil, Nicholas Barton, Marian Vanhaeren, Francesco D'Errico, Simon Collcutt, Tom Higham, Edward Hodge, Simon Parfitt, Edward Rhodes, and Jean- Luc Schwenninger. 2007. 82,000-Year-old shell beads from North Africa and implications for the origins of modern human behavior. *Proceedings of the National Academy of Sciences of the United States of America* 104: 9964–9969.

D'Errico, Francesco. 2003. The invisible frontier: A multiple species model for the origin of behavioral modernity. *Evolutionary Anthropology* 12: 188–202.

Gould, Stephen Jay. 2002. *The structure of evolutionary theory.* Cambridge, MA: Harvard University Press.

Guthrie, R. Dale. 2006. *The nature of Paleolithic art.* Chicago: University of Chicago Press.

Knight, Chris. 1995. *Blood relations: Menstruation and the origins of culture.* New Haven, CT: Yale University Press.

Langdon, John H. 1997. Umbrella hypotheses and parsimony in human evolution: A critique of the aquatic ape hypothesis. *Journal of Human Evolution* 33: 479–494.

McBrearty, Sally, and Alison S. Brooks. 2000. The revolution that wasn't: A new interpretation of the origin of modern human behavior. *Journal of Human Evolution* 39: 453–563.

Mellars, Paul, and Chris Stringer. 1989. Introduction. In Paul Mellars and Chris Stringer, eds., *The human revolution: Behavioural and biological perspectives on the origins of modern humans,* 1–14. Princeton, NJ: Princeton University Press.

Roebroeks, William, Mark J. Sier, Trine K. Nielsen, Dimitri De Loecker, Josep M. Parés, Charles E. S. Arps, and Herman J. Mücher. 2012. Use of red ochre by early Neandertals. *Proceedings of the National Academy of Sciences of the United States of America* 109: 1889–1894.

Soressi, Marie, and Francesco D'Errico. 2007. Pigments, gravures, parures: Les comportements symboliques controversés des Néandertaliens. In B. Vandermeersch and B. Maureille, eds., *Les Néandertaliens: Biologie et cultures,* 297–309. Paris: Éditions du CTHS.

Speth, John D. 2004. News flash: Negative evidence convicts Neanderthals of gross mental incompetence. *World Archaeology* 36: 519–526.

Texier, Pierre-Jean, Guillaume Porraz, John Parkington, Jean-Phillipe Rigaud, Cedric Poggenpoel, Christopher Miller, Chantal Tribolo, Caroline Cartwright, Aude Coudenneau, Richard Klein, Teresa Steele, and Christine Verna. 2010. A Howiesons Poort tradition of engraving ostrich eggshell containers dated to 60,000 years ago at Diepkloof Rock Shelter, South Africa. *Proceedings of the National Academy of Sciences of the United States of America* 107: 6180–6185.

Whewell, William H. 1840. *The philosophy of the inductive sciences, founded upon their history.* London: J.W. Parker.

———. 1858. *History of the inductive sciences from the earliest to the present time,* vol. 2, 3rd ed. New York: D. Appleton.

Wilson, Edward O. 1998. *Consilience: The unity of knowledge.* New York: Knopf.

Wreschner, Ernst E. 1980. Red ochre and human evolution: A case for discussion. *Current Anthropology* 21: 631–644.

Zilhão, João., Diego E. Angelucci, Ernestina Badal-García, Francesco D'Errico, Floréal Daniel, Laure Dayet, Katerina Douka, Thomas F. G. Higham, Maria José

Martínez-Sánchez, Ricardo Montes-Bernárdez, Sonia Murcia-Mascarós, Carmen Pérez-Sirvent, Clodoaldo Roidán-Garcia, Marian Vanhaeren, Valentin Villaverde, Rachel Wood, and Josefina Zapata. 2010. Symbolic use of marine shells and mineral pigments by Iberian Neandertals. *Proceedings of the National Academy of Sciences of the United States of America* 107: 1023–1028.

7 MARK-MAKING AS A HUMAN BEHAVIOR

Ellen Dissanayake

The impressive animal images of the Ice Age, painted and engraved on the walls of deep caves in what are now France and Spain, were discovered a century or so ago. Once this imagery was accepted as genuine, which in some cases took as much as half a century, it was commonly regarded as the "beginning of art"—called that by its discoverers and a public who were inheritors of a Western European elitist idea of *beaux arts*, or fine arts. The images could be dated to the Magdalenian period of the Upper Paleolithic, around 15,000 to 18,000 years ago, and, in 1994, when another spectacular site with painted beasts was discovered—Grotte Chauvet—art's putative origins were pushed back to around 40,000 years ago.

This "paleoart" is usually considered to be a result of ritual or ceremony and, until very recently (and even yet in some quarters), has been taken as evidence for a "cultural revolution" in human cognitive ability that occurred around 40,000 to 50,000 years ago (the boundary between the Middle and Upper Paleolithic). This period was considered a watershed or pivot point in human evolution when humans became cognitively modern and able to symbolize, so that in a "creative explosion" (Pfeiffer 1982) they invented language, art, and religion (Davidson and Noble 1989; Klein 1989; White 1992; Mithen 1996; Noble and Davidson 1996; Wade 2006; Balter 2009).

Many Ice Age depictions are unquestionably magnificent and, as far as we know, unlike anything that came before. Yet if they were sites for ritual behavior, as the powerful images have led many (such as film director Werner Herzog) to believe, it is strange there is no soot on the walls and ceilings of the chambers, because burning torches would have been essential for any sustained activity. Its absence indicates that ceremonies did not take place frequently, if at all. The images may indeed have expressed religious ideas or emotions, but no one can say for sure. They certainly do occasion responses in modern humans that many call "spiritual."

In recent decades, the assumption that art or symbolization began with the Ice Age has been challenged with the discovery of much earlier marks—petroglyphs (literally, "rock carvings")—on stone surfaces, not only in Western Europe, but also in Asia and Africa. These early glyphs do not depict anything recognizable and may occur inside caves, under rock overhangs, or on boulders in plain air—as indeed may have been the case for paintings or pictographs that have weathered away. The lack of identifiable subject matter hinders speculation about their purpose and meaning, although in fact we also have no way of really knowing what figurative renderings meant to those who made or looked at them. Chronology does not help much, because direct dating of any petroglyph is exceedingly difficult and usually impossible. Marks made with charcoal, on the other hand, or paintings with pigment that contains organic binders such as animal protein, blood, eggs, plant resin, honey, or even urine, can produce reliable results by means of radiocarbon dating.

Although early petroglyphs and pictographs are usually included under the broad term *rock art*, they are not uniformly considered "art"—a label that is reserved, at least by many prehistorians, for representational images. They are often described by what they are not: nonrepresentational, noniconic, nonreferential, nonfigurative, or sometimes "geometric" or "abstract." Here, I shall generally call them noniconic, keeping to the original meaning of the word *iconic*—"of (the nature of) an image or portrait," even though the word has been adopted as a descriptive term in semiotics where an icon is a kind of sign—not necessarily a likeness—that may refer to something in the world. In contemporary paleoarchaeology, noniconic marks may be assumed, easily and typically, to be "symbolic," even when it is by no means clear what they are symbolic of.

Controversy about whether noniconic rock art is "art" or "symbolic" can be avoided altogether if we take an ethological rather than analytic and interpretive stance. That is, let us instead think of these images—noniconic as well as iconic—not only as objects to be interpreted and labeled, but also as the tangible and visible residue of a fundamental universal *behavior of mark-making*. Broader, deeper, and equally tantalizing questions can then be asked. Because children spontaneously (without being taught) start to make marks as soon as they can control a drawing instrument manually, one can wonder why this ability should appear universally in our immature descendants as well as in our remote ancestors. Apart from being (or not being) symbolic or art, what do ancient petroglyphs or pictographs share with the Ice Age menageries? What motivated their making? Can we suggest they were adaptive? In other words, did they contribute to the survival and reproductive success of those who made them?

When examining such questions, I use a consilient approach based on one of the axioms of evolutionary and behavioral biology (ethology): universal behaviors of living creatures have evolved and are modified, over time, to adapt them to their way of life. That is, early mark-making can be viewed as part of our ancestors' behavioral tool kit in their lives as hunter-gatherers. In the sections that follow, I describe the earliest known marks made by our Pleistocene ancestors and compare them with marks made by children as they learn to draw. My ideas about mark-making have also been influenced by findings from anthropology, archaeology, child development, and neuroscience. This chapter suggests how a consilient approach can provide new thinking about an ancient and widespread human behavior—mark-making—that heretofore has not even been specifically identified, much less examined, by evolutionary theorists.

The Earliest Rock Markings: Cupules and Incised Striations

It is not widely known that the earliest marks everywhere are noniconic, as is the great majority of all surviving Pleistocene rock art worldwide—including that in Western Europe (Bahn 1998; Sheridan 2005; Watson 2009). The remarkable European cave paintings now seem to be an anomaly, a rare oddity, in a long trajectory of human markings, large and small, on stone surfaces. Despite their abundance and ubiquity, noniconic marks have been largely ignored by the public and overlooked by all but a few scholars. They lack the "mystique" of their more celebrated Ice Age "rock star" cousins, which are unquestionably symbolic and by any standard "art." Although many noniconic marks are beautiful and are carved with great skill, other examples are merely scratched in a casual, perfunctory way and are insignificant in size. (Representational markings, themselves, are also frequently unremarkable.) Yet, even the most ancient and common examples of rock art required diligence and accuracy in their making.

These earliest examples are called *cupules*, referring to their small, roughly hemispherical, concave ("cup"-like) shape. They are found on every inhabited continent, from pre-*sapiens* times, as early as 200,000 years ago at Sai Island in Sudan (Van Peer et al. 2003). They may require from hundreds to many thousands of blows (with a stone tool called a *hammerstone*), depending on the hardness of the stone matrix. Archeologists trying to replicate them report that some of the earliest examples have the smallest possible diameter (32 × 31.5 mm) with the greatest possible depth (9 mm), thus providing evidence of planning, precision, persistence, and considerable skill (Kumar 1996; Kumar and Krishna, 2014).

FIGURE 7.1 Cupules, Daraki-Chattan, Madhya Pradesh, India, more than 170,000 years ago. Photo: Robert Bednarik. Courtesy of Giriraj Kumar, Rock Art Society of India and Robert Bednarik, Rock Art Research.

Cupules are found on horizontal, sloping, and vertical surfaces, often in groups, sometimes arranged in rows. At Daraki-Chattan in present-day Madhya Pradesh, India, two vertical panels are densely covered by 496 cupules of great antiquity (Kumar 1996), predating by perhaps as much as four to five times the paintings or engravings in Ice Age caves in Europe (Figure 7.1.)

Even older than any known cupules is a bone fragment with incised parallel slanted and convergent lines found at Bilzingsleben in Germany, dated to roughly 350,000 years ago (Mania and Mania 1988). The maker was *Homo erectus*, a hominin species that preceded *sapiens* and flourished from roughly 1.8 million years ago to 300,000 years ago. The regularity of the lines suggests they are not accidental cut-marks, but their purpose, if any, is unknown (Figure 7.2). In 2014, a deeply carved, cross-hatched mark made by a Neanderthal, more than 39,000 years ago, was discovered in Gorham's Cave on Gibraltar (Rodriguez-Vidal et al. 2014). A zigzag incision on a 47,000 year-old bone fragment was recovered from the Mousterian shelter of Bacho Kiro in Bulgaria, another Neanderthal site (Bahn and Vertut 1997: 25).

In southern Africa, at rich archaeological sites such as Blombos, engraved markings on small pieces of ochre are being discovered from earlier and earlier times, as ongoing excavations uncover deeper layers. The most recent of these discoveries to date (that is, the earliest artifacts found at the deepest levels so

(a)

(b)

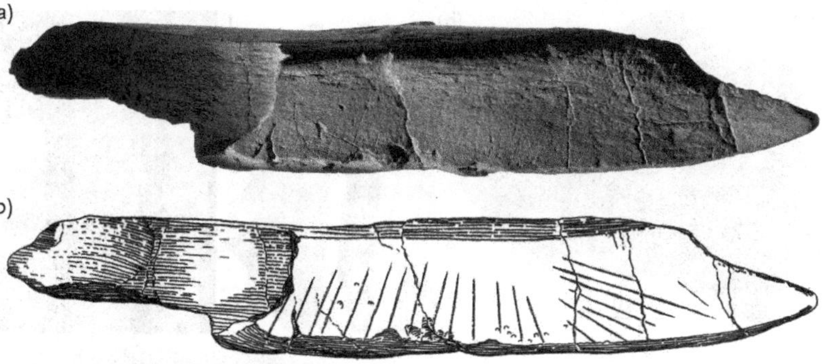

FIGURE 7.2 (a, b) Incised elephant tibia, 350,000 years ago. Photo: Juraj Lipták. Courtesy of Landesamt für Denkmalpflege and Archäologie Sachsen-Anhalt (State Office for Heritage Management and Archaeology Saxony-Anhalt), Halle, Germany.

far, at 99,000 years ago) show parallel, subparallel, vertical, oblique, and convergent (fanlike) striations (Henshilwood et al. 2002) (Figure 7.3). These precede the famous Blombos "Zigzag" of 77,000 years ago, a shaped piece of ochre, 2 × 1.7 inches, on which appears to be intentionally incised[1] a sequence of X marks confined above and below between parallel horizontal bands—a design seen in aboriginal decorative contexts from many places throughout time and to the present day (Figure 7.4). The earliest zigzag yet found was scratched by a *Homo erectus* on a clam shell found recently in Java and has been dated to an astonishing 540,000 to 430,000 years ago, the oldest hominin mark anywhere (Joordens et al. 2015).

In another South African site, Diepkloof rock shelter, 270 fragments from approximately 25 broken ostrich-eggshell containers show engraved repeated abstract patterns that include single and double straight parallel lines, some cross-hatched at right or oblique angles (Texier et al. 2010). (Figure 7.5).

The most recent estimate for the first colonizing of the New World is 16,000 years ago.[2] Although at this time representational rock art was being made in Ice Age Europe and elsewhere, the earliest rock art markings in the Americas (called "Paleoindian" and "Archaic") are, like the earliest marks everywhere, noniconic. At the Gault site in Texas, incised markings on small pieces of limestone or chert, different from but reminiscent of the early linear markings at Blombos, date from as early as 13,000 years ago (Wernecke and Collins 2012) (Figure 7.6).

As far as is known, cupules and deeply grooved lines are believed to be the oldest rock art in the Americas as well as elsewhere (Parkman 1995, 24; 2007). Some occur in groups on single large boulders as well as on vertical, slanting, and horizontal rock outcroppings, as elsewhere in the world (Figure 7.7).

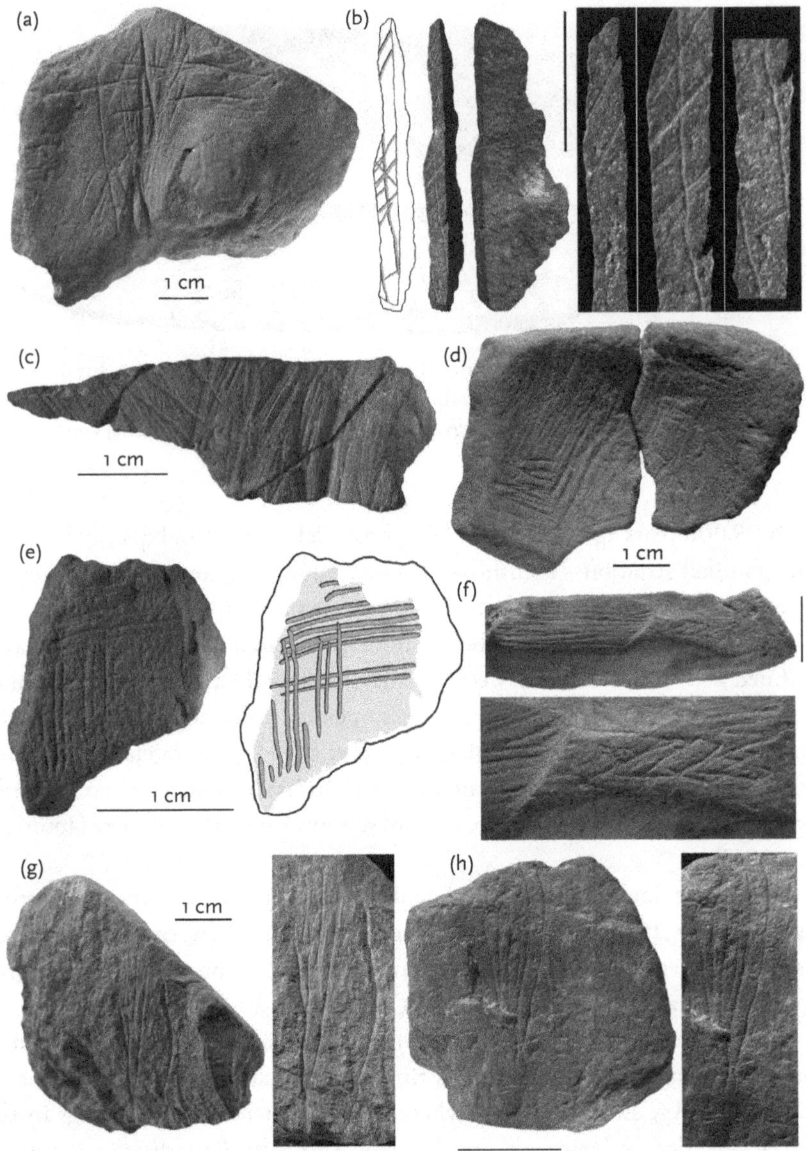

FIGURE 7.3 (a–h) Engravings on ocher, Blombos, South Africa, 99,000 years ago. Courtesy of Professor Christopher Henshilwood.

FIGURE 7.4 Engraving on ocher, Blombos, South Africa, 77,000 years ago. Courtesy of Professor Christopher Henshilwood.

FIGURE 7.5 Decorative engravings on ostrich eggshell fragments, Diepkloof, South Africa, more than 60,000 years ago. Courtesy of Professor Pierre-Jean Texier.

FIGURE 7.6 (a, b) Two engraved stones, ca. 13,000 years ago. Courtesy of The Gault School of Archaeological Research, Texas.

FIGURE 7.7 Arizona cupule site. Photo: Ekkehart Malotki. Courtesy of the Photographer.

The Development of Mark-Making in Children

The development of mark-making in children is curiously similar to that of the human species. As any parent quickly observes, young humans from their first months are preoccupied with using their hands. First they reach out, then grasp and manipulate whatever is close by. This "hand fixation" should not be surprising, for as toolmakers and users humans evolved to find satisfaction and pleasure in using their flexible and dexterous hands.

From birth, hand use develops in a predictable sequence, in which mark-making is a late acquisition. It takes practice and skill to pick up objects and then hold, release, and eventually place, pair, match, sort, order in sequence, and build with them. In developmentally delayed children, these actions may not happen spontaneously, but can be instilled in their natural sequence with help from a trained therapist who introduces the skills in developmental order: grasping, picking up, holding, placing, piling, banging or hammering, scraping, and drawing (Stroh et al. 2008). All these elementary actions are of course used when adults make marks on earth, sand, clay, or stone.

If provided at around age two with a drawing instrument and drawing surface, children will begin to scribble, making large, dynamic quasi-circular movements with the arm and whole hand that grasps the marker; only later, with the achievement of a precision grip, can more controlled drawing begin. The earliest marks made on a suitable surface are based on sensorimotor feelings; they are exploratory and pleasurable, a kind of object play. The movements soon become more controlled, staying within the paper's limits, and the lines become centered and clustered (Fein 1993).

Scribbles (including dots, vertical, horizontal, diagonal, and curved lines, loops, and zigzags) gradually resolve into crossed lines, parallel lines, ladders, grids, and meanders, and eventually become more and more identifiably "geometric" and balanced—variations and combinations of circles, concentric circles, starbursts, labyrinthine forms, and spirals (Figure 7.8).

Children discover these geometric marks spontaneously between ages three and five, and eventually use them as the fundamental elements of their first drawings of humans and animals (Fein 1993; Burrill 2010). Representational form emerges eventually from only four modalities: perpendicular, parallel and oblique lines, and the circle.

Psychologists report that children learn to draw with "orderly growing complexity" (Fein 1993). At first they are motivated to draw, not what they see, but to follow an "inner imperative" (Kellogg 1970; Fein 1993), a self-propelled impetus to make a mark and then follow where it leads, to "play with form" (Alland 1983). In other words, the making itself (including the physical movement and its frequently unforeseen results) is the meaning, without the intent to represent or symbolize.

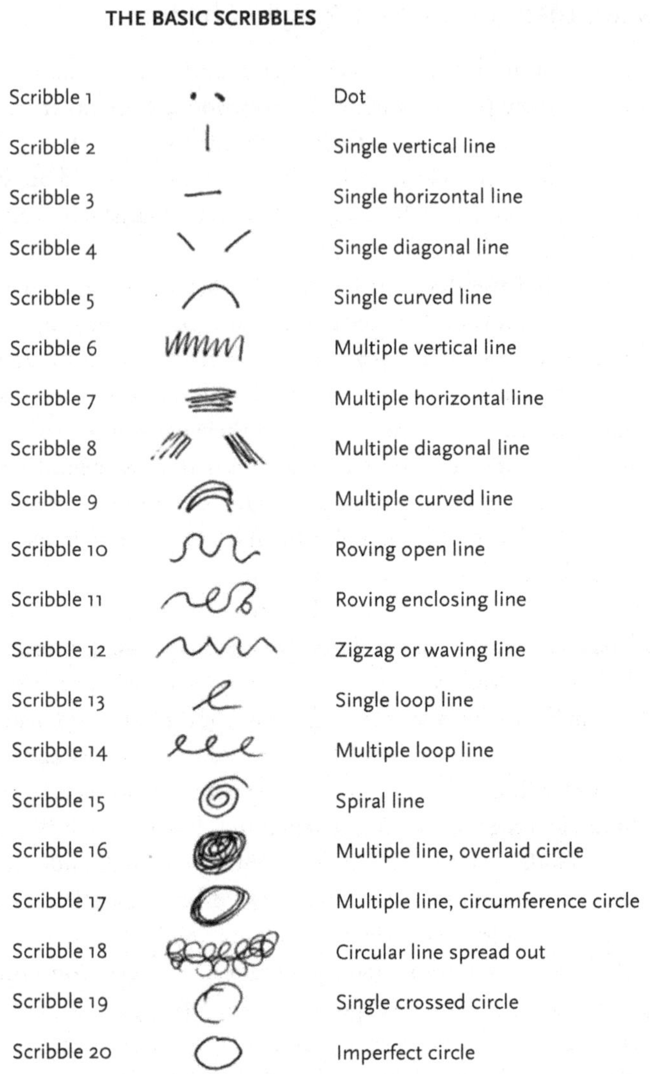

FIGURE 7.8 Children's basic scribbles.
Source: Kellogg, Rhoda. 1970. *Analyzing children's art*, p. 15. Palo Alto, CA: National Press Books.

A Possible Origin of Mark-Making

If we think of origins, there must have been graphic precursors to making marks on stone, rather as complex singing and dancing must have come from simpler kinds of movement and vocalization. From what antecedents could the idea have come?

Recent articles by Sreenathan et al. (2008) and Bednarik and Sreenathan (2012) suggest a new direction for thinking about the origin of mark-making. These authors, and anthropologist Vishvajit Pandya (2009), describe the Jarawa, one of three remaining, small indigenous tribes in the Andaman Islands of the Indian Ocean. All lines of evidence—social, cultural, historical, archaeological, linguistic, phenotypic, and genetic—support the conclusion that the Andaman Islanders as a group were isolated for a substantial period of time and might thus represent an ancient substratum of humanity in Asia, predating later migrations and agrarian expansion events (Endicott et al. 2003). Although the majority of Andamanese people and their languages are long dead, their founding populations may have arrived on the islands in the late Pleistocene—that is, 17,000 to 32,000 years ago (Endicott et al. 2003, 182).

The Jarawa, along with the Onge and Sentinelese, can be differentiated from other native Andaman Islanders (encountered by British seafarers in 1858) by sociocultural, phenotypic, and linguistic delineations, and by geographic distribution (Endicott et al. 2003) (Figure 7.9). Although the Jarawa have been somewhat acquainted with modern technological society since 1999, for the most part they have maintained a traditional hunter-forager-fisher economy, unlike the Onge, who have sustained rapid cultural loss since the 1960s (Pandya 2009, 157). The Sentinelese, sole residents of North Sentinel Island, remain isolated and hostile to this day (Pandya 2009, 144) and have never been the subject of ethnographic study.

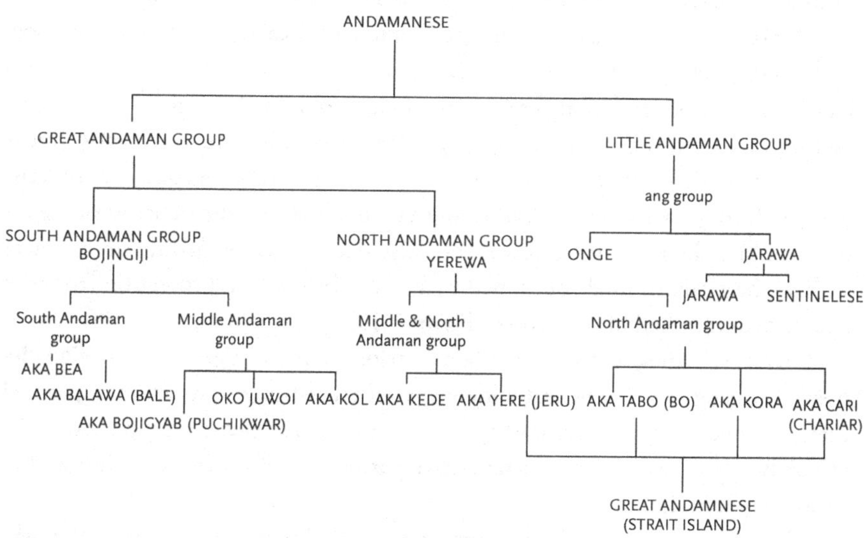

FIGURE 7.9 Diagram of Andamanese tribal divisions. Diagram: Robert Bednarik. © Robert Bednarik, M. Sreenathan, and *Rock Art Research*.

This history suggests that the cultural practices of the Jarawa may have developed in isolation from the influences of other tribal populations (except for the Onge and Sentinelese, with whom they share some cultural features) for as long as 17,000 to 32,000 years, as just described, or even ca. 50,000 to 70,000 years ago (Thangaraj et al. 2002; Thangaraj et al. 2005). This is not to claim that every part of their culture has remained unchanged since that time, but simply to say that because the groups have been isolated from other influences for a far longer time than most other contemporary tribal populations known to anthropologists, one might conjecture that at least some of their cultural practices may provide a close-to-precontact view of one group of hunter-gatherers.[3] In particular, Jarawa body decoration seems worth describing here insofar as it supports a suggestion that the practice of making geometric designs on the body is ancient—arguably occurring even in the Pleistocene—and could have been a source for the idea and practice of making marks on other surfaces, whether soft clay or hard stone. Jarawa body markings are typically composed of elementary linear patterns that are casually or even carelessly applied by individuals to themselves. More elaborate work is accomplished with the help of others, especially wives painting husbands.

It is common knowledge that people who live in small-scale societies in warm climates need few, if any, garments. Yet even when nearly or completely unclothed, they frequently ornament their bodies with selected shells, leaves, flowers, and fruits; and animal fur, teeth, or bones. In ceremonial contexts, when they sing, dance, and give dramatic performances, they decorate themselves even more elaborately. In many such groups, people may also paint their bodies and some possessions with what can be called graphic "art" that is exclusively noniconic.

The Jarawa are no exception to these practices, wearing ornaments of bone, shell, and woven natural fiber as well as plants, flowers, and—recently—headbands made of shredded and repurposed cloth obtained as "gifts" from "contact parties" and, now, tourists (Pandya 2009, 104, 108, 146). They also smear white clay on their face and body and then scrape a variety of geometric patterns with their fingernails and fingers. Using a mixture of clay and ochre with pig fat or water and the juice of a creeper, they also decorate a few utilitarian artifacts such as chest-guards made of shaped bark, worn by males as protection from the release of a stretched hunting bow (Figure 7.10).

On faces, bodies, and some artifacts, various noniconic geometric design elements—zigzags, narrowly and widely spaced cross-hatches, fish bones, parallel lines, lozenges, loops, and small circles—occur both as elementary patterns and in combinations as continuous repetitive patterns enclosed between horizontal bands (Figure 7.11).

Although these styles and patterns have names, they are conventional, without further assigned meaning.[4] There is no suggestion of an origin in nature for

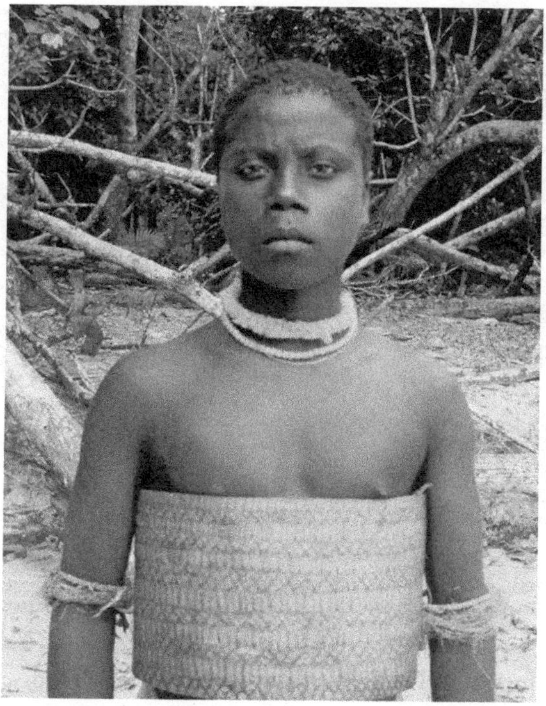

FIGURE 7.10 Jarawa-decorated chest guard. Photo: M. Sreenathan. © Robert Bednarik, M. Sreenathan, and *Rock Art Research*.

any mark. All members of the Jarawa community recognize the designs. It is interesting that the same design elements are found widely in other parts of the world from Blombos and Diepkloof to rock markings in North America. Some are seen also in children's early mark-making. Parallel lines, traveling zigzags and traveling loops, used by Jarawa, are common marks made by two- to three-year-old children (Matthews 2013, 341–342) (Figure 7.12).

How might the idea of decorating the body have arisen? Like other small-scale societies, the Jarawa treat a variety of ailments, including fever, swellings, or injuries on their bodies, by applying plant parts sometimes mixed with red and white clay (Figure 7.13). Could graphic body decoration have developed from such a practice? One can imagine hominin ancestors applying mud to their bodies for cooling or for protection from insects or the sun, as well as for healing.

Seeing the textures of finger marks or ridges on the skin could well suggest the deliberate making of such marks with fingers in clay, earth, sand, or even on stone (Klemm 1989 [1951]; Sreenathan and Bednarik 2008, 381).[5] Finger marks naturally create straight parallel lines, both horizontal and vertical, or meanders. I suggest that it is possible that such marks eventually might have been used

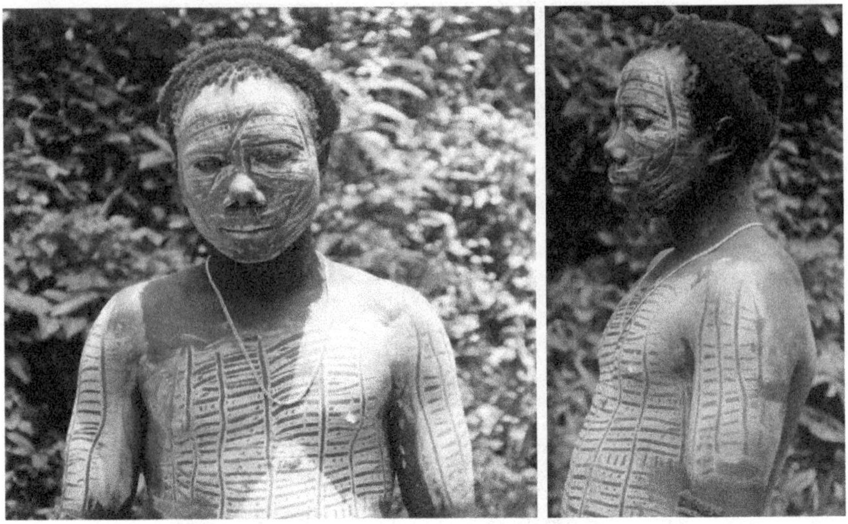

FIGURE 7.11 Two views of a Jarawa man with body painting. Photo: M. Sreenathan. © Robert Bednarik, M. Sreenathan, and *Rock Art Research*.

deliberately in formalized and repeated ways on bodies and possessions, perhaps to mark people as "human"—just as insects (such as butterflies or beetles), reptiles, fish, and mammals were observed to have their recognizable characteristic markings. Like many tribal people, the Jarawa call themselves their word for "human": *əng*. Although speculative, I find it plausible that the idea could have arisen of displaying this state visually.

A further speculation is that, long before the Ice Age, it was juvenile male humans (inspired by noniconic body-painting designs) who first began to make marks on sandy or clay surfaces, then on stone, in play or to show off for comrades or girls. This is not as unlikely as it might at first appear. By age 14 or 15, at least, Pleistocene males would be essentially as strong as they would ever be, but still juvenile in behavior and psychological maturity (or immaturity). Although they would be acquiring male skills from childhood, they would not be required, at least for a period of time, to take on adult responsibilities. Something similar has been suggested by Dale Guthrie (2005), who makes a plausible case that some painters of images in Ice Age caves were adolescents (see also Bednarik 2008a; Bednarik and Sreenathan 2012). Adolescent males would have had the risk-taking propensity, physical agility and strength, motivation, and leisure to explore caves, fool around, talk about animals and female bodies (the predominant subject matter of the cave images), and draw them. Guthrie himself is a hunter and recalls his own early hunting trips with age-mates. In addition, he made a careful study of

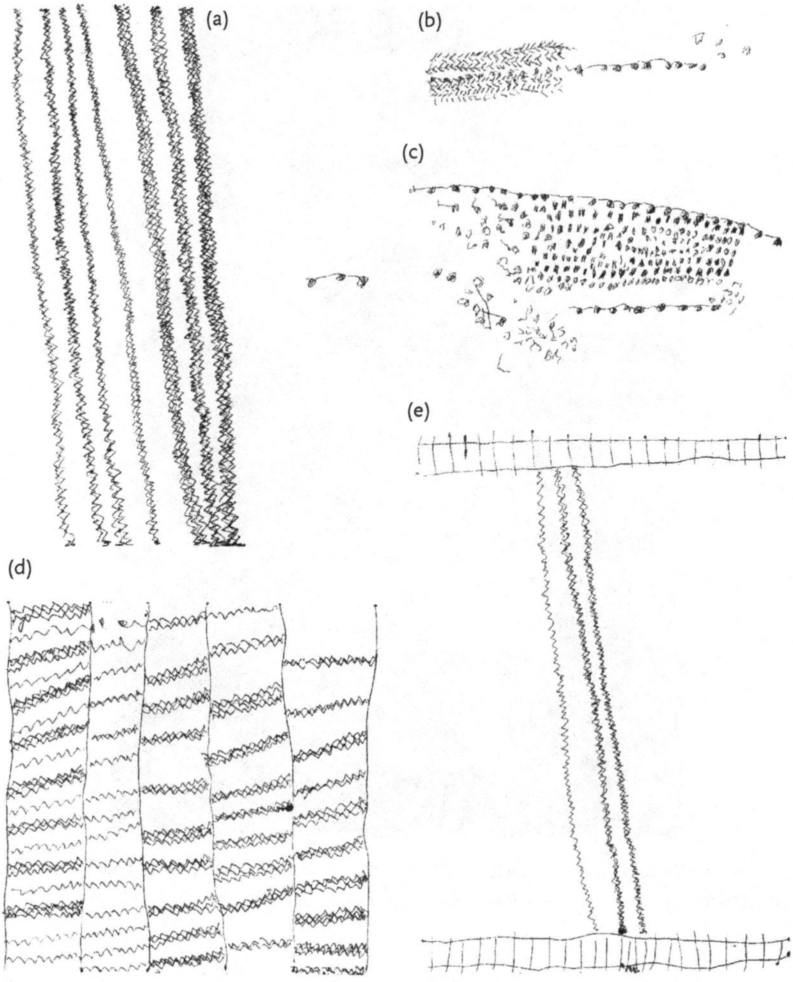

FIGURE 7.12 (a–e) Jarawa design motifs. Drawing: Robert Bednarik. © Robert Bednarik, M. Sreenathan, and *Rock Art Research.*

almost every handprint in Ice Age caves and compared them with prints of his own students at the University of Alaska, concluding they were made by young people, both male and female.

When mark-making became established as a cultural practice, markings could eventually become associated with (a "sign" or even "symbol" of) an idea, appropriated by adult males, and at some point become the focus of ritual observance. Although we may never know, I agree with Bednarik and Sreenathan (2012) and Hodgson (2014) that it is a reasonable hypothesis.

FIGURE 7.13 Jarawa boy with body painting. Photo: M. Sreenathan. © Robert Bednarik, M. Sreenathan, and *Rock Art Research*.

What Motivated Mark-Making?

It is much easier to describe markings that people make, what they look like, than to suggest why humans made them—that is, why (or whether) they were adaptive. Evolutionists distinguish between "proximate" and "ultimate" causes of human behavior. With respect to behavior, proximate causes involve the motivations for behaving in a certain way, and the neural and physiological mechanisms that correspond to motives. For example, one feels hungry (neural and physiological stimulation produces the sensation of hunger) and the resulting motivation is to seek food. The proximate benefit of the behavior is to reduce the unpleasant feeling of hunger and its ultimate adaptive effect is maintaining the body and thus survival.

The adaptive reasons for cultural actions such as making marks, or making a certain kind of mark, are not so straightforward, although mark-making itself appears to be predisposed biologically. Like other primates, humans evolved to have dexterous and manipulative hands that accomplish many adaptive things for their lives, so it is not surprising that it is satisfying, both physically and psychologically, to use these hands with precision and skill. Even so, making cupules or parallel grooves on stone does not satisfy an obvious adaptive need, such as finding and ingesting food. And although everyone has to eat, not everyone feels an irresistible need to make permanent marks on rock surfaces, despite the millions that have been made during thousands of millennia and across many thousands of miles. It is also important to recognize that an individual's motivation for behaving in a certain way, even for making a mark, may not be adaptive at all. And in the case of cultural behaviors, there are intervening variables such as cultural belief systems that motivate behavior; mark-making is not usually just an appetitive urge. With these points in mind, let us look at some of the many proximate reasons that have been suggested as motivations for mark-making.

Some marks were probably simply accidental, a side effect of grinding, in the case of some cupules; or the unintended residue of cutting something else, as in some early linear marks. In addition, marks may sometimes have been straightforwardly utilitarian, such as (with some apparent cupules) making a depression to hold water, medicine, or a food substance. Some early marks could have served didactic purposes as knapping guidelines (Althin 1950; Fischer [1974] in Wernecke and Collins [2012, CD 677]); game boards (Bednarik 2008b, 31); teaching aids, or aides-memoire (Marshack 1972); a trail marker for way-finding; a tally or record (noting time or quantity; handprints showing "I was here"); a territory marker (as with a clan sign or symbol); or an indication of status (as with acquiring a tattoo or cicatrice, a person may be required to make a cupule or other glyph to indicate a cultural achievement such as attaining adulthood or becoming a warrior).[6]

Some mark-making on soft surfaces quite possibly resulted from idle doodling (Watson 2008) or was simply a by-product of pleasure in hand use—a kind of play or pastime (Koch-Grünberg [1907], as cited in Boas [1927, 118]). Certainly, those among our ancestors who did not enjoy using their hands to make things for their life would not have survived as well as their more handy companions whose offspring would have inherited the trait for finding pleasure and meaning in making (Dissanayake 1995).

Yet any one of these suggestions does not apply universally for every mark. Marks apparently have a wide variety of possible and actual uses—some perhaps adaptive, others most likely not. Mark-making on stone remains an evolutionary puzzle, even more than other early-appearing arts, because it is especially laborious, with much less apparent immediate reward than singing, dancing, or watching dramatic performances.

My own suggestion is that once mark-making was in place, originating in exploratory and playful hand use, as described earlier, it became an adjunct to ritual behavior, which is adaptive (Dissanayake 1992; Rappaport 1999; Wade 2009; Dissanayake 2012). All human societies perform ceremonial practices or rituals in which several arts combine—song, dance, and dramatic storytelling, in addition to the visual panoply of costume and other body adornment, masks, altered surroundings, and special objects, which could include painted or carved marks on stone. It is not known when our ancestors began to engage in ceremonial practices, but it is possible, even likely, that at least some early rock markings are at former sites for these activities and are today their only vestiges.

In some instances, perhaps the act of mark-making itself, and its side effects, was the proximate motivation. In one example recorded in 1940, an Aboriginal Australian man pounding a cupule at Tukalili (Northern Territory) told an ethnographer that the resulting fine mineral powder was pink cockatoo sperm that was released to fertilize the site as a kind of life essence (Mountford 1976). In some cases, the repetitive movement of making a cupule or other mark may have predisposed the maker to an altered state of consciousness, leading to a vision (Parkman 1993, 99).

It is also possible that the motivation for making a mark was auditory—to provide a sound (Ouzman 2005). Such sounds could be intended to effect a natural occurrence (such as persuading the wind to change direction) or as a summons for people to gather at a particular place or to be aware that "something [important] is going to happen here" or "is happening here."

In fact, any cupule or mark could become an intentional or unintentional record that "something [important] happened here" and then a focus for later participation and enactment, perhaps with further marks added at a later time or times. It is possible that simply the act of creating a rock art marking may have been of primary importance (Steinbring and Buchner 1997, 77). Some of the early engraved portable stones from Gault, Texas, were deliberately broken after their manufacture, leading their finders to suggest that the process and the patterns made were more important than the finished product (Wernecke and Collins 2012, CD 671).

Was Mark-Making Adaptive?

Scholars have suggested ultimate adaptive advantages of the arts of dancing, singing, and performing, which could perhaps also apply to mark-making. For example, the so-called "sexual selection" hypothesis proposes that by demonstrating personal strength, skill, and creativity, an individual's status is enhanced, thereby attracting mates (Miller 2000, 2001; Varella et al. 2011). The sexual selection hypothesis seems plausible until it is pointed out that it is applicable to almost any activity: indeed, it is really about virtuosity, creativity, and other descriptors of special ability in any domain, and does not necessarily explain what would be adaptive about that domain in the first place, apart from providing an opportunity to show off. In the specific case of mark-making, although marks may draw a viewer's attention, not all marks are particularly distinctive or highly skilled, and in any case few could be traceable to one especially superior alpha-maker.

The "coalition-signaling" hypothesis (Hagen and Bryant 2003) suggests that group performances of vigorous music and dance could display the prestige and power of a social group, thereby impressing (and dominating) others. Although the sight of numerous petroglyphs in one place might be considered intimidating, the hypothesis seems less likely to apply to static marks than to the immediate performance activities for which the hypothesis was originally proposed.

As with the various suggestions for proximate motivations described earlier, these two hypotheses are too broad and unspecific to give a widespread ultimate adaptive benefit to mark-making. When any behavior has evolved, it can be used for display, but many adaptive behaviors, such as tool-making, language, or play, are better explained as predispositions that give adaptive benefits to all humans, not only the best craftspeople, orators, or athletes—or the best mark-makers.

Neuroscientist Derek Hodgson (2000) suggests that an adaptive benefit of mark-making is its contribution to our perceptual organization of the real world. Graphic "primitives" such as lines (giving contour and edge), parallel lines, zigzags, grids, and various closed forms that use angles (Hodgson 2007) reflect the cognitive architecture of the visual system of the brain. Making and responding to such marks fine-tunes neurons that are necessary for the rapid discrimination of things in the world, thereby increasing awareness of environmental signals by helping to sharpen and hone important perceptual skills (Hodgson 2006, 31). Such "default aesthetic biases" (Palmer et al. 2012) simulate as well as stimulate the processes by which the visual system constructs form from primitives (Hodgson 2006, 32).

By privileging passive visual perception, Hodgson's "neurovisual resonance theory" overlooks the role of movement in art-making (Burrill 2010) and the motor neural constraints on *making* geometric marks. Lines, meanders, and circles mimic the types of continuous movements our arms can make, as shown in children's earliest mark-making, when the crayon doesn't leave the drawing surface. Making repeated dots, with the hand knocking the crayon against the surface, is another kind of motor movement that provides practice and the pleasure of having an effect of the world. From a purely kinesiological perspective, geometrics are easier to create than figurative art (Steven Brown, pers. comm. March 12, 2009).

The neurovisual resonance hypothesis receives support from a contrast between Pleistocene and modern environments. Pleistocene environments were not "carpentered," as are the built environments that, in the modern world, offer omnipresent horizontals, verticals, diagonals, and other geometric designs. In a Pleistocene environment, human-made marks would be distinctive. However, other primates develop perceptual acuity to deal with their visual and spatial worlds without additional exercise of their sensory apparatus.

Both visual– and motor–neuroscientific suggestions for a behavior of adult mark-making seem compatible with what we know about the earliest mark-making of children. But apart from exploratory and pleasure-based play or a sense of developing mastery and control, they don't suggest a proximate reason (after childhood) for an adult to want to continue to make marks, especially on intractable stone. Few adults make marks in order to exercise and satisfy their neurovisual perceptual or motor pathways, although pleasure, satisfaction, and even improved pattern recognition may result.

Neither suggested source for mark-making proposes a specific ultimate benefit. As with other abilities that are developed in play—whether motor, social, linguistic, or imaginative—the impetus for pleasurable mark-making can be co-opted and used in a wide array of other contexts, some adaptive, others not.

Mark-Making and Ceremonial Rituals

My own view of the ultimate adaptive benefits of mark-making differs from the hypotheses described earlier, even as I concur that one or another might apply in specific instances. As mentioned in the previous section, proximate motivations for mark-making, as for other artlike behaviors, seem likely to have been associated with ceremonial ritual or religious practice. In effect, a ceremony is inseparable from its arts. It is not too much to say that, without the arts (song, percussion, dance, costume, visual panoply, site enhancement), there is no ceremony. Their costliness (the amount of time, energy, and material resources they consume) indicates arts are not superficial accessories, but necessities.

Ceremonial rituals, wherever they occur, are intended to affect outcomes of important but uncertain biological needs—for food, safety in the hunt or in battle, health, prosperity, fertility, and successfully passing through inevitable and crucial life transitions such as birth, puberty, marriage, adulthood, and death (Malinowski 1954 [1925]). The artlike components of ritual, just described, are a means to influence spirits to assist in obtaining desired ends or preventing unwelcome ones. Regardless of whether a positive outcome for a particular ceremony is achieved (the proximate motivation for performing it), at least two physiological and neurological ultimate adaptive effects are accomplished.

First, simply participating with others in a time-honored practice that is expected to have positive results reduces anxiety (as measured by stress hormones), leading to better physical and psychological health of individual members of the group. Second, ceremonial participation contributes to group unification and one-heartedness, ensuring, at least for a time, that individuals will work together in confidence and unity. These two effects of ceremonies are accomplished by means of active participation in the ongoing temporal organization and forward movement of the ceremony—through dance and song, which entrain (coordinate and unify) individuals neurologically, physically, emotionally, and psychologically. Opioids such as oxytocin are released in the brain during rhythmic, repetitive activity with other persons, instilling feelings of elation, trust, and bondedness, and these brain chemicals also reduce the pernicious effects of stress-induced cortisol (Carter 1998; Nelson and Panksepp 1998; Uvnäs-Moberg, 1998; Carter and Altemus 1999; Miller and Rodgers 2001; Heinrichs et al. 2003; Taylor et al. 2008; Dissanayake 2012; Dunbar et al. 2012).

However, making permanent marks on rock by pecking, chiseling, or engraving with a stone tool is different from ceremonial participation. Although, like song and dance, it requires time and physical effort, the activity is usually solitary; it not only lacks the positive effects of social participation, but also it exposes the mark-maker to the curiosity of animal or human predators. It is difficult to see an adaptive advantage to such apparently pointless and costly behavior, which depletes energy and time that could be devoted to something more beneficial, such as finding or preparing food, courting sexual partners, making tools and weapons, or even just resting.

Nonetheless, insofar as mark-making contributes to the overall effect of a ceremony (such as contributing a regular, repetitive sound or simply being a prescribed part of an emotionally charged panoply of varied physical activities and sensory impressions), one might consider it adaptive *as an instance of the broader behavioral predisposition* that I call *artification* (see next section). Insofar as individuals grasp and respond to the purpose and meaning of a mark, it would contribute to group unification and the social order. Such marks may serve as a kind of emotional communication, part of a shared intense experience that imbues the image with power. Because the spot remains as a record of an important event or transfiguring idea, later observers will feel the spot to be special, having meaning for themselves and their fellows—as occurs with memorial markers such as the Vietnam Veterans Memorial in Washington, DC (Dissanayake 2012).

A Consilient Concept: Artification

Recall the "creative explosion" hypothesis described at the beginning of this chapter: around 50,000 years ago, cognitively modern humans (suddenly) developed the ability to symbolize and invented religion, language, and art. Ancient noniconic rock art markings challenge that once axiomatic assumption and support a "gradualist" position that the development of symbolic expression occurred over a much longer period of time.

I offer a hypothesis that augments the gradualist view by positing an evolved behavioral predisposition in ancestral humans to "artify." The term is not a synonym for art or art-making, but refers instead to a prior foundational evolved human capacity to make ordinary reality (in the case of mark-making, an ordinary body or stone surface) extra-ordinary, even if the result may not be beautiful, skilled, symbolic, good, and so forth—the usual characteristics attributed to "art." The Jarawas's body decorations are not particularly skilled or beautiful. In the magnum opus that resulted from his studies of what at the time was called "primitive art," Franz Boas (1955, 12–13) proposed there were two sources of artistic effect: one based on form alone and the other on ideas associated with form. Individual body designs of the Jarawa do not appear to have a "meaning" other than adorning, or perhaps originally showing a simple distinction between humans and other forms of life.

Artifiers use special devices that attract attention, sustain interest, and create and manipulate emotion. Such devices include simplification or formalization, repetition, exaggeration, elaboration, and manipulation of expectation. The ethological term for such devices or "operations," when used instinctively by other animals in visual, vocal, and movement modalities, is *ritualized behaviors* (and the process is called *ritualization*), reflecting their formal, repetitive, exaggerated nature (e.g., Huxley 1914; Tinbergen 1952; Eibl-Eibesfeldt 1971; Lorenz 1982). It is not surprising that humans also are predisposed to use them deliberately and respond to them instantly, often with pleasure.

These devices of artification might be called "*aesthetic* primitives" (Dissanayake 2015), immediately attracting attention in any sensory modality (rather as, but in a manner different from, lines, contours, edges, and other visual primitives that automatically stimulate the visual cortex) because they are recognized perceptually, cognitively, and emotionally as being unlike ordinary or familiar stimuli. Their effect can be momentary or sustained, of mild interest or overpowering affect. Perceptions that startle or dazzle manipulate expectation and provide emotions of surprise, wonder, fear, and awe. They seem to occupy another order of experience or state of being that can easily be interpreted as "spiritual."

But responses to artifications can be other than aesthetic. At the least, any artification indicates intentionality; someone has deliberately made or done something. Because they are not accidental, other humans who perceive or experience them infer they "mean" something: artification is a sign of importance (Sütterlin 2003)—whether it be subject matter, the occasion, the consequences of the occasion, or generally all three. Moreover, if we find evidence of artifications from the past—that is, if we recognize aesthetic primitives in artifacts or at a site—we can assume they marked something that many people, or at least one person, cared about. Although artifications today have many manifestations and purposes, and are often contentious and propagandistic, it is still the case that people artify things they care about. Humans are typically not content to leave ordinary reality alone, especially in circumstances about which they care greatly—weddings, graduations, inaugurations, and local or national catastrophes, to name a few.

It is important to emphasize that the process of (and responses to) artification and its aesthetic primitives may well be nonverbal or presymbolic. This is obvious in ritualized behaviors of animals and even in the drawings of children, whose earliest marks are not intended by them to represent anything. Just as music and dance may, but need not, symbolize anything, so human mark-making is not automatically symbolic. In fact, just as artification precedes symbolizing ability in our children, it probably did so in our species. To put it another way, rather than conceptualizing art as a subset of a presumed symbolizing ability that is required for making a drawing or painting, I suggest that symbolic marks and actions are a subset of the larger, and prior, category of artification.

A symbolic mark, like any mark, can itself be artified or used to artify something else. But a symbol itself need not be artified; although, when it is important, it usually is. The symbolic words on this page are not artified, although the words of the title on the jacket cover of the book are because, importantly, they are meant to attract the attention of potential readers (buyers).

The consilient ethological approach used in this chapter identifies mark-making as an evolved behavioral predisposition that includes diverse forms of paleoart: cupules, incised lines, and geometric shapes. Mark-making is here regarded as a type or kind of artification. I identify similarities of form and motive in early human mark-making and in the mark-making of young children. Focusing on *noniconic* examples of mark-making rather than on representational images that suggest religious or other symbolism, and using an ethnographic study of noniconic body decoration in a contemporary aboriginal group, I aim at putting into question entrenched archaeological and anthropological assumptions about "art." I cast doubt on art's automatic connection with seriousness, symbolization, spirituality, modern cognition, and other presumed evidence of a "creative explosion."

I do not disagree that much of human art is serious, symbolic, religious, and cognitively complex. But a consilient study of archaic marks makes clear there are other impulses inspiring artlike behavior and it is unwise to assume we can understand our remote ancestors as having "cognitively modern" minds, if that means they were like ours today. To the consilient view presented here, painting or engraving animal images on the walls of caves and other rock surfaces is a late-appearing instance or version of an underlying universal impetus to make marks, which itself can be regarded as part of an even deeper human universal—making ordinary "reality" extra-ordinary.

Acknowledgments

I am grateful to Marialuisa Ciminelli, Ken and Ann Gary, and Ekkehart Malotki for helpful comments on an earlier version of this chapter, and to Julia and Patrick Malotki for assistance with Figure 7.8.

I am ever grateful to Professor Ekkehart Malotki for introducing me to rock art study and for reading and commenting on this essay.

Notes

1. A stimulating study by Matteo Scardovelli (2013) examined the zigzag artifact (Blombos 1–6) and found that the marks producing the zigzag pattern were not sequentially, but haphazardly, incised, suggesting that a preconceived pattern was not in the mind of the maker. See also Malotki and Dissanayake (manuscript in preparation) for further discussion of this artifact.
2. Most scholars now believe that people first entered the New World through the Bering land bridge more than 16,000 years ago. After entering Alaska, it is not known whether they colonized the hemisphere by moving down the Pacific coast, by inland routes, or both. The general view is that early immigrants would have spread down the coast much faster than they could move inland because they could exploit familiar coastal resources more readily and get much of their food from the sea. However, evidence to support the coastal migration theory has been particularly hard to find because sea levels at the time were about 200 feet lower than they are today. As the sea level rose, it covered most of the early coastal settlements (Dillehay et al. 2008; Curry 2012).
3. The first modern European contact with the islands was in 1780; various visitations, mainly by the British and Indians, took place thereafter. In 1958, a protected forest reserve of 765 km^2 (a fraction of their former tribal land) was set up for the Jarawa

and, until 1999, they were one of the last remaining Andaman Island tribes not to have been assimilated into the neighboring colonial society—especially those Jarawa who lived in the forest interior and the east coast of their territory (Pandya 2009, 107, 206, 207). The entire native Andamanese population declined after early contact from approximately 6500 to 485 in the year 2000 (Pandya 2009, 74).

4. Hodgson (2014, 65) suggests the opposite sequence—that the neurovisual appeal of geometric marks may have been later "exploited for personal/individual decorative use," as in body decoration, which was then "exapted for wider socio-cultural purposes involving group affiliation."
5. In the Onge, some applications of white clay paint on the body are associated with clan identity and with the individual's season of birth, or to mark recent ritual transitions from states such as mourning or illness (Pandya 2009, 110). Calling these uses "symbolic" of these states does not seem warranted (see Malotki and Dissanayake, manuscript in preparation).
6. A useful annotated list of these and other suggested functions for abstract/geometric markings on rock has been compiled by Wernecke and Collins (2012).

References

Alland, Alexander, Jr. 1983. *Playing with form*. New York: Columbia University Press.

Althin, Carl-Axel. 1950. New finds of Mesolithic art in Scania (Sweden). *Acta Archaeologica 21*: 253–260.

Bahn, Paul G. 1998. *The Cambridge illustrated history of prehistoric art*. Cambridge: Cambridge University Press.

Bahn, Paul G. and Jean Vertut 1997. *Journey through the ice age*. Berkeley: University of California Press.

Balter, Michael 2009. On the origin of art and symbolism. *Science* 323: 709–711.

Bednarik, Robert G. 2008a. Children as Pleistocene artists. *Rock Art Research* 25: 173–182.

———. 2008b. Cupules. *Rock Art Research* 25: 61–100.

Bednarik, R. G., and M. Sreenathan. 2012. Traces of the ancients: Ethnographic vestiges of Pleistocene "art." *Rock Art Research* 29: 191–217.

Boas, Franz 1955 [1927]. *Primitive art*. New York: Dover.

Burrill, Rebecca. 2010. The primacy of movement in art making. *Teaching Artist Journal* 8: 216–228.

Carter, C. Sue 1998. Neuroendocrine perspectives on social attachment and love. *Psychoneuroendocrinology* 23: 779–818.

Carter, C. Sue, and Margaret Altemus. 1999. Integrative functions of lactational hormones in social behavior and stress management. In C. Sue Carter, I. Izja Lederhendler, and Brian Kirkpatrick, eds., *The integrative neurobiology of affiliation*, 361–371. Cambridge, MA: MIT Press.

Curry, Andrew. 2012. Ancient migration: Coming to America. *Nature* 485: 30–32.

Davidson, Iain, and William Noble. 1989. The archaeology of perception: Traces of depiction and of language. *Cambridge Archaeological Journal* 30: 125–155.

Dillehay, Tom D., C. Ramirez, M. Pino, Michael B. Collins, Jack Rossen, and J. D. Pino-Navarro. 2008. Monte Verde: Seaweed, food, medicine, and the peopling of South America. *Science* 320: 784–786.

Dissanayake, Ellen 1992. *Homo aestheticus: Where art comes from and why.* New York: Free Press.

———. 1995. The pleasure and meaning of making. *American Craft* 55: 40–45.

———. 2012. The deep structure of Pleistocene rock art: The "artification hypothesis." In Clottes, J. (dir.). L'art plèistocéne dans le monde [Pleistocene Art of the World], Actes du Congrés IFRAO [International Federation of Rock Art Organizations] September 2010, Symposium 9, Signs, Symbols, Myth, Ideology. Special Issue of *Préhistoire, Art et Sociétés,* Bulletin of the *Socièté Préhistorique Ariège-Pyrénées, LXV-LXVI, 2010-2011,* p. 278, CD: 1601–1611.

———. 2015. "Aesthetic primitives": Fundamental biological elements of a naturalistic aesthetics. *Aisthesis: Pratiche, Linguiaggi e Saperi dell'Estetico* 8: 6–24.

Dunbar, Robin I. M., Kostos Kaskatis, Ian MacDonald, and Vinnie Barra. 2012. Performance of music elevates pain threshold and positive affect: Implications for the evolutionary function of music. *Evolutionary Psychology* 10: 688–702.

Eibl-Eibesfeldt, Irenäus. 1971. *Love and hate: The natural history of behavior patterns,* trans. Geoffrey Strachan. New York: Holt, Rinehart and Winston.

Endicott, Phillip, M. Gilbert, P. Thomas, Chris Stringer, Carles Lalueza-Fix, Eske Willersley, Anders J. Hansen, and Alan Cooper. 2003. The genetic origins of the Andaman Islanders. *American Journal of Human Genetics* 72: 178–184.

Fein, Sylvia. 1993. *First drawings: Genesis of visual thinking.* Pleasant Hill, CA: Exelrod.

Fischer, Anders. 1974. An ornamented flint-core from Holmegård V, Zealand, Denmark. *Acta Archaeologica* 45:155–168.

Guthrie, R. Dale. 2005. *The nature of Paleolithic art.* Chicago: University of Chicago Press.

Hagen, Edward H., and Gregory A. Bryant. 2003. Music and dance as a coalition signaling system. *Human Nature* 14: 21–51.

Heinrichs, Markus, Thomas Baumgartner, Clemens Kirschbaum, and Ulrike Ehlert. 2003. Social support and oxytocin interact to suppress cortisol and subjective responses to psychosocial stress. *Biological Psychiatry* 54: 1389–1398.

Henshilwood. Christopher S., Francesco d'Errico, Royden Yates, Zenobia Jacobs, Chantal Tribolo, Geoff A. T. Duller, Norbert Mercier, Judith C. Sealy, Helene Valladas, and Ian Watts. 2002. Emergence of modern human behavior: Middle Stone Age engravings from South Africa. *Science* 295: 1278–1280.

Hodgson, Derek. 2000. Art, perception and information processing: An evolutionary perspective. *Rock Art Research* 17: 3–34.

———. 2006. Altered states of consciousness and palaeoart: An alternative neurovisual explanation. *Cambridge Archaeological Journal* 16: 27–37.

———. 2007. The earliest manifestations of "art": An attempted integration. In Peddarapu Chenna Reddy, ed., *Exploring the mind of ancient man: Festschrift to Robert G. Bednarik*, 25–34. New Delhi: Research India Press.

———. 2014. Decoding the Blombos engravings, shell beads and Diepkloof ostrich eggshell patterns. *Cambridge Archaeological Journal* 24: 57–69.

Huxley, Julian. 1914. The courtship habits of the great crested grebe (*Podiceps cristatus*) together with a discussion of the evolution of courtship in birds. *Journal of the Linnaean Society of London: Zoology* 53: 253–292.

Joordens, Josephine C. A., Francesco d'Errico, Frank P. Wesselingh, Stephen Munro, John de Vos, Jakob Wallinga, Christina Ankjærgaard, Tony Reimann, Jan R. Wijbrans, Klaudia F. Kuiper, Herman J. Mücher, Hélène Coqueugniot, Vincent Prié, Ineke Joosten, Bertil van Os, Anne S. Schulp, Michel Panuel, Victoria van der Haas, Wim Lustenhouwer, John J. G. Reijmer & Wil Roebroeks 2015. *Homo erectus* at Trinil on Java used shells for food production and engraving. *Nature* 518: 228–231.

Kellogg, Rhoda. 1970. *Analyzing children's art.* Palo Alto, CA: National Press Books.

Klein, Richard G. 1989. Biological and behavioural perspectives on modern human origins in Southern Africa. In Paul Mellars, and Chris Stringer, eds. *The human revolution,* 530–546. Edinburgh: Edinburgh University Press.

Klemm, Gustav. 1989 [1851]. *The four elements of architecture and other writings*, trans. Harry Mallgrave and Wolfgang Herrmann. New York: Cambridge University Press.

Kumar, Giriraj. 1996. Daraki-Chattan: A Palaeolithic cupule site in India. *Rock Art Research* 13: 38–46.

Kumar, Giriraj, and Ram Krishna. 2014. Understanding the technology of the Daraki-Chattan cupules: The cupule replication project. *Rock Art Research* 31: 177–186.

Lorenz, Konrad Z. 1982. *The foundations of ethology: The principal ideas and discoveries in animal behavior*, trans. Konrad Z. Lorenz and Robert Warren Kickert. New York: Simon and Schuster.

Malinowski, Bronislaw. 1954 [1925]. *Magic, science, and religion.* Garden City, NY: Doubleday Anchor Books.

Mania, Dietrich, and Ursula Mania. 1988. Deliberate engravings on bone artefacts of *Homo erectus*. *Rock Art Research* 5: 91–107.

Marshack, Alexander. 1972. *The roots of civilization.* London: Weidenfeld and Nicolson.

Matthews, John. 2013. Seven spots and a squiggle: The prehistory of pictures. In Klaus Sachs-Hombach and Jörg R. J. Schirra, eds. *Origins of pictures,* 311–352. Köln: Halem.

Miller, Geoffrey F. 2000. *The mating mind: How sexual choice shaped the evolution of human nature.* New York: Doubleday.

———. 2001. Aesthetic fitness: How sexual selection shaped artistic virtuosity as a fitness indicator and aesthetic preferences as mate choice criteria. *Bulletin of Psychology and the Arts* 2: 20–25.

Miller, Warren B. and Joseph Lee Rodgers. 2001. *The ontogeny of human bonding systems: Evolutionary origins, neural bases, and psychological mechanisms.* Dordrecht: Kluver.

Mithen, Steven. 1996. *The prehistory of the mind: The cognitive origins of art.* London: Thames and Hudson.

Mountford, Charles P. 1976. *Nomads of the Australian desert.* Adelaide: Rigby.

Nelson, Eric E., and Jaak Panksepp. 1998. Brain substrates of infant–mother attachment: Contributions of opioids, oxytocin, and norepinephrine. *Neuroscience and Biobehavioral Reviews* 22: 437–452.

Noble, William, and Iain Davidson. 1996. *Human evolution, language and mind: A psychological and archaeological enquiry.* Cambridge: Cambridge University Press.

Ouzman, Sven. 2005. Seeing is deceiving: Rock art and the non-visual. In Christopher Chippindale and Paul S. C. Taçon, eds. *Aesthetics and rock art,* 253–268. London: Ashgate.

Palmer, Stephen. E., Karen B. Schloss, and Jonathan Sammartino. 2012. Hidden knowledge in aesthetic judgments: Preferences for color and spatial composition. In Arthur P. Shimamura and Stephen E. Palmer, eds., *Aesthetic science: Connecting minds, brains, and experience,* 189–222. Oxford: Oxford University Press.

Pandya, Vishvajit. 2009. *In the forest: Visual and material worlds of Andamanese history (1858–2006).* Lanham, MD: University Press of America.

Parkman, E. Breck. 1993. Creating thunder: The Western rain-making process. *Journal of California and Great Basin Anthropology* 15: 90–110.

———. 1995. A theoretical consideration of the pitted boulder. *Papers on Californian Prehistory* 4: 1–24.

______. 2007. Pit-and-groove antiquity in new world prehistory. *Science Notes* 89: 1–23. Petaluma, CA: California State Parks.

Pfeiffer, John E. 1982. *The creative explosion: An inquiry into the origins of art and religion.* New York: Harper and Row.

Rappaport, Roy A. 1999. *Ritual and religion in the making of humanity.* New York: Cambridge University Press.

Rodriguez-Vidal, J., F. d'Errico, F. G. Pacheco, R. Blasco, J. Rosell, R. P. Jennings, A. Queffelec, G. Finlayson, D. A. Fa, J. M. Lopez Gutiérrez, J. S. Carrión, J. J. Negro, S. Finlayson, L. M. Cáceres, M. A. Bernal, S. F. Jiménez, and C. Finlayson. 2014. A rock engraving made by Neanderthals in Gibraltar. *PNAS Early Edition* www.pnas.org/cgi/doi/10.173/pnas.1411529111 [Online].

Scardovelli, Matteo. 2013. The myth of the "symbolical" Blombos Cave piece of ochre. Paper presented at the IFRAO International Rock Art Conference. Albuquerque, New Mexico, May 26–31.

Sheridan, Susan Rich. 2005. A theory of marks and mind: The effect of notational systems on hominid brain evolution and child development with an emphasis on exchanges between mothers and children. *Medical Hypotheses* 64: 417–427.

Sreenathan, M., V. R. Rao, and Robert G. Bednarik. 2008. Paleolithic cognitive inheritance in aesthetic behavior of the Jarawas of the Andaman Islands. *Anthropos: International Review of Anthropology and Linguistics* 103: 367–392.

Steinbring, Jack., and A. P. Buchner. 1997. Cathedrals of prehistory: Rock art sites of the Northern Plain. *American Indian Rock Art* 23: 73–84.

Stroh, Katrin, Thelma Robinson, and Alan Proctor. 2008. *Every child can learn: Using learning tools and play to help children with developmental delay.* London: Sage.

Sütterlin, Christa. 2003. From sign and schema to iconic representation: Evolutionary aesthetics of pictorial art. In Eckart Voland and Karl Grammer, eds., *Evolutionary aesthetics*, 131–170. Berlin: Springer.

Taylor, Shelley E., Lisa J. Burklund, Naomi I. Eisenberger, Barbara J. Lehman, Clayson J. Hilmeet, and Matthew D. Lieberman. 2008. Neural bases of moderation of cortisol stress responses by psychosocial resources. *Journal of Personality and Social Psychology* 95: 197–211.

Texier, Pierre. Jean, Guillaume Porraz, John Parkington, Jean-Philippe Rigaud, Cedric Poggenpoel, Christopher Miller, Chantal Tribolo, Caroline Cartwright, Aude Coudenneau, Richard Klein, Teresa Steele, and Christine Verna. 2010. A Howiesons Poort tradition of engraving ostrich eggshell containers dated to 60,000 years ago at Diepkloof Rock Shelter, South Africa. *Proceedings of the National Academy of Science of the United States of America* 107: 6180–6185.

Thangaraj, Kumarasamy., Lalji Singh, Alla G. Reddy, V. Raghavendra Rao, Subhash C. Sehgal, Peter A. Underhill, Melanie Pierson, Ian G. Frame, and Erika Hagelberg. 2002. Genetic affinities of the Andaman Islanders, a vanishing human population. *Current Biology* 13: 86–93.

Thangaraj, Kumarasamy, Gyaneshwer Chaubey, Toomas Kovisild, Alla G. Reddy, Vijay Kumar Singh, Avinash A. Rasalkar, and Lalji Singh. 2005. Reconstructing the origin of Andaman Islanders. *Science* 308: 996.

Tinbergen, Niko. 1952. "Derived" activities: Their causation, biological significance, origin and emancipation during evolution. *Quarterly Review of Biology* 17: 1–32.

Uvnäs-Moberg, Kerstin. 1998. Oxytocin may mediate the benefits of positive social interaction and emotions. *Psychoneuroendocrinology* 23: 819–835.

Van Peer, Philip, Richard Fullagar, Stephen Stokes, Richard M. Bailey, Jan Moeyersons, Frans Steenhoudt, Aurora Geerts, Tim Vanderbeken, Morgan De Dapper, and Francis Geus. 2003. The Early to Middle Stone Age transition and the emergence of modern human behaviour at site B-11, Sai Island, Sudan. *Journal of Human Evolution* 45(2):187–193.

Varella, Marco Antônio Corrêa, Altay Alves Lino de Souza, and José Henrique Benedetti Piccoli Ferreira. 2011. Evolutionary aesthetics and sexual selection in the evolution of rock art aesthetics. *Rock Art Research* 28: 153–186.

Wade, Nicholas. 2006. *Before the dawn: Recovering the lost history of our ancestors.* New York: Penguin Press.

———. 2009. *The faith instinct: How religion evolved and why it endures.* New York: Penguin Press.

Watson, Benjamin. 2008. Oodles of doodles? Doodling behavior and its implications for understanding paleoarts [with comments]. *Rock Art Research* 25: 35–60.

———. 2009. *Universal visions: Neuroscience and recurrent characteristics of world palaeoart.* PhD diss., University of Melbourne, Australia.

Wernecke, D. Clark, and Michael B. Collins. 2012. Patterns and process: Some thoughts on the incised stones from the Gault Site, Central Texas, United States. In IFRAO [International Federation of Rock Art Organizations] Congress 2010. Symposium 3: Pleistocene Art of the Americas. CD: 669–681.

White, Randall. 1992. Beyond art: Toward an understanding of the origins of material representation in Europe. *Annual Review of Anthropology* 21: 537–564.

IV INTEGRATIVE PSYCHOLOGY

8 CONSILIENCE THROUGH THE INTEGRATION OF ENGINEERING AND SOCIAL SCIENCE

Barbara Oakley

A Specific Example of How an Engineering Approach Added Insight to Social Science Research

Optical illusions can provide extraordinary insights into how the brain works. One of the most subtle and provocative is that created by Roger Shepard, "Turning the Tables" (Shepard 1990). In Figure 8.1, modeled after Shepard's work, the table on the left seems longer and thinner than the one on the right. In reality, however, the two table tops are precisely the same size and shape—the table top is simply rotated at a different angle. The illusion occurs because of the apparent—but not actual—expansion of space with distance. The illusion is so compelling that even when a paper is cut and placed onto one tabletop, and then moved to the other, the exact fit seems like magic.

This illusion works because our ability to perceive objects in three-dimensional space is automatic and deeply entrenched in our neurological systems. With no conscious effort whatsoever, we draw our conclusions. As Shepard (1990) notes:

> [W]e cannot choose to see a drawing merely as what it is—a pattern of lines on a flat, two-dimensional surface. . . . that pattern automatically triggers the circuits in the brain that make the three-dimensional interpretation appropriate to such a perspective display. Any consciously adopted intentions to ignore such an interpretation are largely powerless against the swift deliverances of this underlying machinery. (126–127)

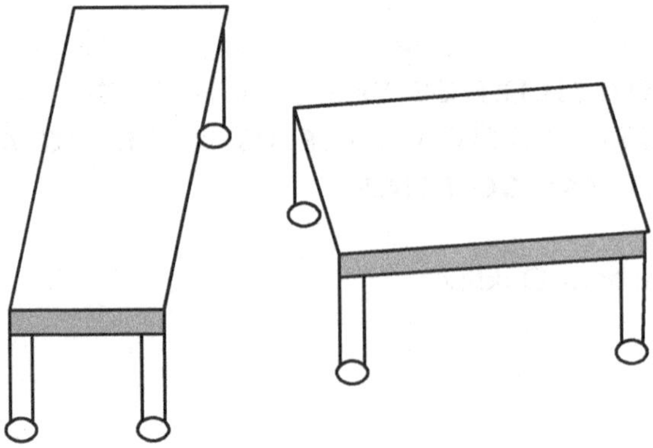

FIGURE 8.1 Turning the tables.

As it occasions, our empathetic feelings for others, coupled with a desire to be liked, and on occasion, narcissism (we feel we know what is best for others [Post 1993]), can lead us into illusions of helping—*pathological altruism* (Oakley et al. 2012). Much as with visual illusions, illusions of helping arise from deep automatic triggers in the brain that sometimes render individuals "largely powerless against the swift deliverances of this underlying machinery" (Shepard 1990, 127).

How might we address this type of challenge? One way is through learning to use a pragmatic, engineering approach. In fact, the central idea of this chapter is that bringing insights from engineering into the world of social sciences can be an innovative way to gain new and useful perspectives that can uncover new avenues of research.

In pathological altruism, an agent's underlying motivation is to promote another's welfare and, as a consequence, to feel good about his or her own actions. But instead of overall beneficial outcomes, this type of "altruism" can have irrational and substantial negative consequences to others or even to the self. An example would be the mother who thinks she is helping her obese child by giving him the cookies he craves. The evidence seems utterly convincing to this well-meaning mother that she is doing what is best for her child; she uses motivated reasoning to discount medical arguments to the contrary.

Other examples of pathological altruism include pet hoarders, who truly believe they are helping the animals they rescue, even when they are stepping over the dead bodies of animals that have died in their care (Nathanson and Patronek 2012). Serial adopters of children, it seems, may sometimes be less motivated by proper care of their charges than by the saintly treatment they receive from their in-group for their ostensive sacrifice. Codependent women who find it difficult to say no to another's needs ultimately appear to suffer far higher than normal rates of depression and burnout as a consequence of their overly helpful behavior (McGrath and Oakley 2012). Would-be philanthropists who donate to support nongovernmental agencies can, in a sad twist, turn out to be important supporters of sadistic totalitarian dictators (Easterly 2006; Moyo 2009; Polman 2010). In the ultimate example of pathological altruism, genocidal killers who truly believe they are helping their own in-group, exterminate "vermin" who happen to be other human beings (Brannigan 2012). All of these, of course, provide extreme cases. However, there are many gradations, along with a myriad of other more pronounced examples.

Much as with optical illusions, illusions of helping grow from our automatic and deeply entrenched emotional responses involving empathetic caring for others. As Kahneman describes in *Thinking, Fast and Slow*, and as many neuroscientific studies have revealed, these caring feelings for others occur at a subconscious level, and can be very difficult to set aside (Mathur et al. 2010; Shamay-Tsoory 2010; Decety 2011a, 2011b; Kahneman 2011; Breithaupt 2012a, 2012b; de Waal 2012).

Perhaps surprisingly, although altruism has received substantive attention in the research literature, pathologies of altruism, and harmful biases that arise because of altruistic intent, have received very little attention. (Oakley et al., 20 12, Oakley, 2013) My work in social science research, as described in this chapter, was strongly motivated by my training in engineering. Specifically, my engineering training has brought a strong understanding of the importance of trade-offs. Engineering, although frequently seen as simplistic and deterministic by those who are unfamiliar with the discipline, is a nuanced field that trains individuals to realize there is often no ideal solution to any specific challenge—and that it is important to understand the trade-offs of any seemingly beneficial program or design.

Moreover, engineering is also a discipline that demands a strong scientific background, and this, too, has informed my approach to examining altruism. A central element of engineering is that it is vitally important to test and prove theorems and designs from many perspectives before accepting their validity. In the social sciences, on the other hand, there is a rich tradition of problematic concepts and theories being broadly accepted for many decades with little attention paid even to flagrant counterfactuals (see, for example, Baars 1986, Brannigan 2004, Carnahan and McFarland 2007, Straus 2007, Makel and Plucker 2014, Stoet and Geary 2015). Ultimately, then, I was strongly motivated to look at research into the double-edged helpful–harmful nature of altruism because of my training in engineering.

How Engineering Can More Generally Inform Social Science

Academic approaches and theories have an enormous impact, not only on public perceptions, but also on legislation and policy at both a state and federal level. As such, it is vital that academic approaches and theories be well grounded in provable, factual reality. A concrete example is seen in the dramatic improvement of medical care with the introduction of an evidence-based approach for the past 150 years. We have come a long way from the "obviously beneficial" bloodletting and cupping of the 1800s (Chalmers 2006).

A similar advance is vital for the social sciences. As history reveals, social scientists surprisingly often put forth well-meaning but untested or poorly evaluated programs and theories that inadvertently worsen the very situation they were meant to solve (Cole 2001; Wright and Cummings 2005; Wilson 2011, Chapter 2). This is not an inconsequential problem; in fact, these untested and problematic theories can have serious consequences for society as a whole. Many of these theories and programs can go on for decade after decade at enormous cost, despite benefits that are illusory.

Examples of this phenomenon include such therapies as "critical incident stress debriefing." Not only does this costly therapy not mitigate the symptoms of posttraumatic stress disorder, it may actually worsen them (Rose et al. 2003). Other examples include well-meaning programs to reduce racial prejudice through diversity and affirmative action, and programs to reduce adolescent behavior problems and drug use (Petrosino et al. 2000, 2003; Kalev et al. 2006; Lilienfeld 2007; Eidelson and Soldz 2012). As psychologist Timothy Wilson (2011) notes in *Redirect: The Surprising New Science of Psychological Change*:

> Know some teens headed for trouble? Scare the heck out of them by taking them to prisons and funeral homes. Want kids to avoid drugs? Bring police officers into their classrooms to explain the dangers of drugs and give lessons on how to resist peer pressure. These approaches make perfect sense but, it turns out, they are perfectly wrong, doing more harm than good. It is no exaggeration to say that commonsense interventions have prolonged stress, raised the crime rate, increased drug use, made people unhappy, and even hastened their deaths. (24–25).

The underlying problem is the seductive nature of the program or theory being proposed. Often such efforts sound so obviously beneficial it is difficult to imagine that preliminary testing might be necessary. Even when a proposed program looks questionable on the face of it, it is difficult for critics to bring themselves to criticize it publicly. After all, criticism of the program might imply the critic is against the "good thing" the program is proposing to accomplish (Straus 2007, 2008, 2009; Sunstein 2009).

This scenario is very different from that in engineering.

Both engineering and social science seek to provide benefits to society. Engineering does this by creatively applying scientific principles to the design and development of devices, structures, and processes. Engineering differs from the social sciences in that engineers receive harsh, relatively rapid real-world feedback when a creative application is wrong or contains flaws. Bridges fall, devices fail, processes go awry, and when that happens the results can be devastating (Petroski 1992, 2006). As a consequence, engineers naturally acquire through their training—and perhaps even through their underlying neurological underpinnings (Billington et al. 2007)—a sense that any theory put forth and any device constructed must be checked carefully. All presumptions must be based on sound principles. It's not enough that something sounds good or seems as if it would probably work. That way, in fact, lies danger.

Integration of Engineers and Engineering Approaches into the Social Sciences

Engineering has a unique relationship with the humanities and social sciences. In general, there *isn't* a relationship. In fact, there is a long history of antagonism between the humanities—and often the social sciences—on one side, and science and engineering on the other (see, for example, Snow 1964; Becher and Trowler 2001; Kean 2011.) A glimmer of consilience is beginning to appear as a consequence of the past decade's breakthroughs in neuroscience. Most important, however, opportunities for consilience are arising as evolutionary theory is increasingly applied to our understanding of the humanities and social sciences (Wilson 1975, 1998; Barrett et al. 2002; Wilson 2003; Owen 2007; Carroll 2011).

In the spirit of consilience, it is clear that an engineering mind-set of creative, pragmatic problem-solving skills and approaches could be of great value in the social sciences. Such an engineering–social scientist mind-set could, in fact, help address the fundamental problem of pathological altruism in social science theory and practice.

A fruitful new approach would be to encourage students to consider majoring in the seemingly disparate disciplines of engineering and psychology. One might think that the discipline of "engineering psychology" already provides the proper combination. But engineering psychology, as a branch of industrial and organizational psychology, simply involves the scientific study of employees, workplaces, and organizations. It does not involve training in the critical engineering mind-set and approach. Moreover, there are only approximately 1200 positions nationally for industrial and organizational psychologists—and these positions generally require a doctorate (Bureau of Labor Statistics 2011a). In contrast, there are some 1.4 million engineering positions available nationally—positions that generally require only a bachelor's degree (Bureau of Labor Statistics 2011b); 1.4 million is arrived at by counting the positions currently available nationwide for the engineering disciplines noted here. In other words, providing students with dual training in both psychology and engineering would provide them with many career opportunities that would not be available if they were simply to major in the psychological discipline "engineering psychology."

The ultimate goal of a dual engineering–psychology degree would be the creation of a small pool of engineer–psychologists who could go on to complete their psychology doctoral studies, bringing with them their analytical and creative problem-solving skills. At the same time, some engineer–psychologists could go on to bring their people skills to bear in the engineering workplace—a valuable asset in the modern, team-based working environment.

Inviting engineers to speak at psychology conferences—discussing, for example, creativity—could be another fruitful new method of building bridges between the disciplines. In my own experience, there has long been an unfortunate tendency from those in the humanities and social sciences to disparage the sciences and technology as uncreative. The reality is, however, that engineering creativity in the building and improvement of a myriad of devices such as the refrigerator, automobile, suspension bridge, computer, and the pacemaker have arguably accomplished as much or more to improve and ease the lives of human beings as the creativity of social scientists.

Psychologists, sociologists, and other social scientists who are also trained as engineers could bring a wealth of new ideas and approaches to the social sciences. But perhaps even more important, bringing individuals trained to use a pragmatic rational approach to the social sciences could do much to begin building a better understanding of the importance of pragmatic verification of new theories, no matter how obviously beneficial such theories may sound in their initial stages. This is particularly important in relation to one of the most intractable problems of the social sciences: the illusory sense that something that *sounds* good must be *doing* good.

Again, academic approaches and theories from the social sciences have an enormous impact, not only on public perceptions, but also on legislation and policy at both a state and federal level. There is much to be gained by bringing

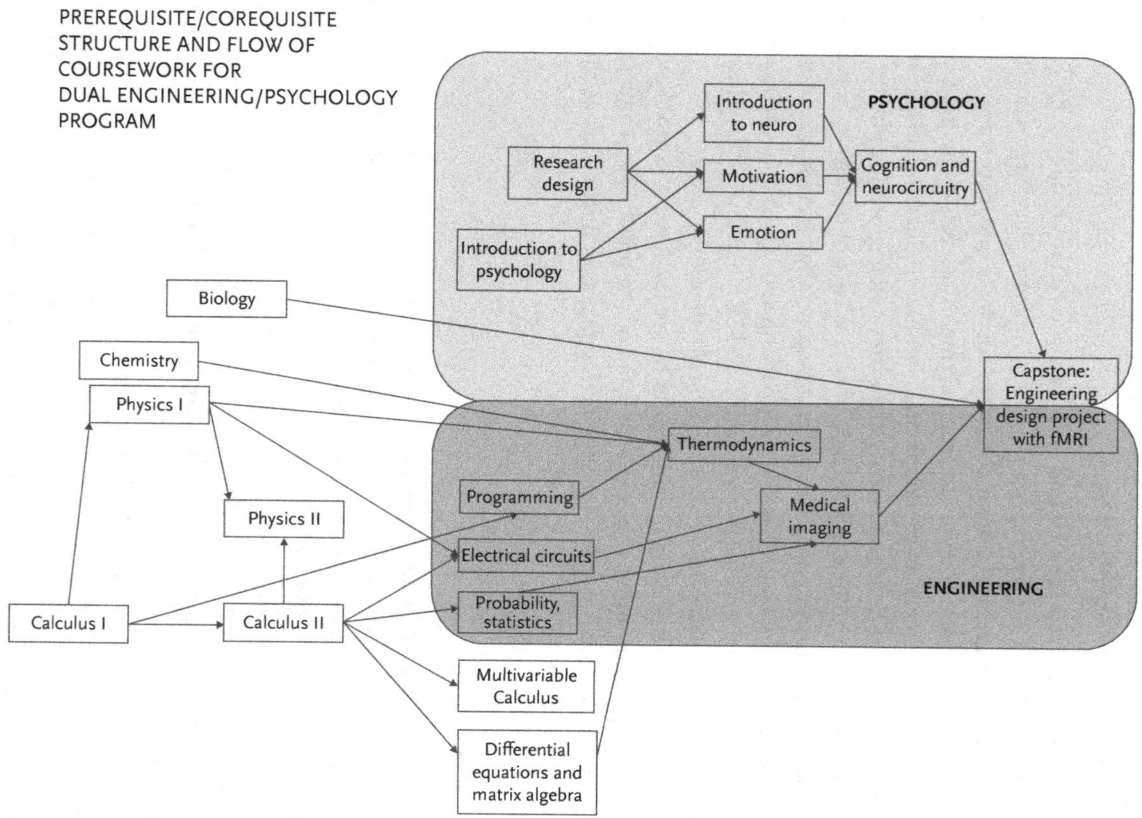

FIGURE 8.2 A potential psychology–engineering program.

the practical, analytic, yet creative approach of engineering to the social sciences, so that the impact of social science in the long run is genuinely for the better of humankind.

Programmatic Suggestions

Engineering programs in the United States are accredited through ABET, the Accreditation Board for Engineering and Technology. There are many types of engineering programs—among the most common and popular are civil, mechanical, electrical, and computer engineering. Much less commonly, there are unaccredited engineering programs that combine engineering with another scientific discipline, such as engineering chemistry and engineering physics.

Programs that combine engineering with the science of biology have proved themselves so useful and popular they are now generally formally accredited—ABET currently has 79 accredited bioengineering and biomedical engineering programs (which have a medical orientation), and 29 biological engineering programs (which don't necessarily have a medical orientation).

A possible course of action, then, would be to develop a new psychology–engineering program using as a template some of the previous bioengineering programs. A potential psychology–engineering program might look as shown in Figure 8.2.

Conclusion

Engineers are trained in the importance of model and theory—and verification of those models and theories through practical exercises. Creating new mechanisms for interdisciplinary interactions between the social sciences and engineering may be a creative new way to assist social scientists in their efforts to help others while simultaneously applying pragmatic approaches grounded on science. Social science models and explanatory mechanisms would, in particular, gain from engineering insight into the development and verification of models and mechanisms.

As with many disciplines, including both medicine and engineering, the focus in the social sciences is on helping others. However, the social sciences have not grown from a pragmatic scientific tradition that seeks solid verification and confirmation of theories before widespread implementation of therapies arising from those theories. Moreover, the impetus for helping others seems to have a blinding effect; testing for "obviously" beneficial therapies is often thought to be unnecessary (Oakley 2013). Ultimately, academia, like many other environments, can lend itself to emotional contagion, so that impractical, misleading, and even harmful ideas that are patently false can take root and become widely believed. Pragmatic engineering approaches, wedded to profoundly beneficial conceptions from the social sciences, could become a powerful force for genuine altruism in improving the human condition.

References

Baars, Bernard J. 1986. *The cognitive revolution in psychology*. New York: Guilford.

Barrett, Louise, Robin Dunbar, and John Lycett Lycett. 2002. *Human evolutionary psychology*. Princeton: Princeton University Press.

Becher, Tony, and Paul Trowler. 2001. *Academic tribes and territories*. Philadelphia: Open University Press.

Billington, Jac, Simon Baron-Cohen, and Sally Wheelwright. 2007. Cognitive style predicts entry into physical sciences and humanities: Questionnaire and performance tests of empathy and systemizing. *Learning and Individual Differences* 17: 260–268.

Brannigan, Augustine. 2004. *The rise and fall of social psychology: The use and misuse of the experimental method*, New York: Aldine de Gruyter.

———. 2012. Genocide: From pathological altruism to pathological obediance. In Barbara Oakley, Ariel Knafo, Guruprasad Madhavan, and David Sloan Wilson, eds., *Pathological altruism*, 225–236. New York: Oxford University Press.

Breithaupt, Fritz. 2012a. A three-person model of empathy. *Emotion Review* 4: 84–91.

———. 2012b. Empathy does provide rational support for decisions. But is it the right decision? *Emotion Review* 4: 96–97.

Bureau of Labor Statistics. 2011a. *Occupational employment and wages, May 2014 19–3032 industrial-organizational psychologists*. http://www.bls.gov/oes/current/oes193032.htm [Online].

———. 2011b. *Occupational outlook handbook: Architecture and engineering occupations*. http://www.bls.gov/ooh/architecture-and-engineering/home.htm [Online].

Carroll, Joseph. 2011. *Reading human nature: Literary Darwinism in theory and practice*. Albany: SUNY Press.

Carnahan, Thomas, and Sam McFarland. 2007. Revisiting the stanford prison experiment: Could participant self-selection have led to the cruelty? *Personality and Social Psychology Bulletin* 33: 603.

Chalmers, Iain. 2006. Why fair tests are needed: A brief history. *Evidence Based Medicine* 11: 67–68.

Cole, Stephen, ed. 2001. *What's wrong with sociology?* New Brunswick, NJ: Transaction.

Decety, Jean. 2011a. Dissecting the neural mechanisms mediating empathy. *Emotion Review* 3: 92–108.

———. 2011b. Promises and challenges of the neurobiological approach to empathy. *Emotion Review* 3: 115–116.

de Waal, Frans. B. 2012. The antiquity of empathy. *Science* 336: 874–876.

Easterly, William. 2006. *The white man's burden: Why the West's efforts to aid the rest have done so much ill and so little good*. New York: Penguin.

Eidelson, Roy, and Stephen Soldz. 2012. *Does comprehensive soldier fitness work: CSF research fails the test*. Coalition for an Ethical Psychology. http://www.ethicalpsychology.org/Eidelson-&-Soldz-CSF_Research_Fails_the_Test.pdf [Online].

Kahneman, Daniel. 2011. *Thinking, fast and slow*. New York: Farrar Straus & Giroux.

Kalev, Alexandra, Frank Dobbin, and Erin Kelly. 2006. Best practices or best guesses? Assessing the efficacy of corporate affirmative action and diversity policies. *American Sociological Review* 71: 589–617.

Kean, Sam. 2011. Red in tooth and claw among the literati. *Science* 332: 654–656.

Lilienfeld, Scott O. 2007. Psychological treatments that cause harm. *Perspectives on Psychological Science* 2: 53–70.

Makel, Matthew C., and Jonathan A. Plucker. 2014. Facts are more important than novelty: Replication in the education sciences. *Educational Researcher* 43: 304–316.

Mathur, Vani A., Tokiko Harada, Trixie Lipke, and Joan Y. Chiao. 2010. Neural basis of extraordinary empathy and altruistic motivation. *NeuroImage* 51: 1468–1475.

McGrath, Michael, and Barbara Oakley. 2012. Codependency and pathological altruism. In Barbara Oakley, Ariel Knafo, Guruprasad Madhavan, and David Sloan Wilson, eds., *Pathological altruism*, 49–74. New York: Oxford University Press.

Moyo, Dambisa. 2009. *Dead aid: Why aid is not working and how there is another way for Africa*. New York: Farrar, Straus and Giroux.

Nathanson, Jane, and Gary Patronek. 2012. Animal hoarding: How the semblance of a benevolent mission becomes actualized as egoism and cruelty. In Barbara Oakley, Ariel Knafo, Guruprasad Madhavan, and David Sloan Wilson, eds., *Pathological altruism,* 107–115. New York: Oxford University Press.

Oakley, Barbara A. 2013. Concepts and implications of altruism bias and pathological altruism. *Proceedings of the National Academy of Sciences of the United States of America* 110: 10408–10415.

Oakley, Barbara, Ariel Knafo, Guruprasad Madhavan, & David Sloan Wilson, eds. 2012. *Pathological altruism*. New York: NY: Oxford University Press.

Owen, Tim. 2007. After postmodernism: Towards an evolutionary sociology. In Jason L. Powell and Tim Owen, eds., *Reconstructing postmodernism: Critical debates,* 153–165. New York: Nova Science.

Petrosino, Anthony, Carolyn Turpin-Petrosino, and James O. Finckenauer. 2000. Well-meaning programs can have harmful effects! Lessons from experiments of programs such as Scared Straight. *Crime & Delinquency* 46: 354–379.

Petrosino, Anthony, Carolyn Turpin-Petrosino, and John Buehler. 2003. Scared Straight and other juvenile awareness programs for preventing juvenile delinquency: A systematic review of the randomized experimental evidence. *The Annals of the American Academy of Political and Social Science* 589: 41–62.

Petroski, Henry. 1992. *To engineer is human: The role of failure in successful design.* New York: Vintage Books, Random House.

———. 2006. *Success through failure: The paradox of design.* Princeton: Princeton University Press.

Polman, Linda. 2010. *The crisis caravan: What's wrong with humanitarian aid?* Trans. Liz Walters. New York: Metropolitan Books.

Post, Jerrold. 1993. Current concepts of the narcissistic personality: Implications for political psychology. *Political Psychology* 14: 99–121.

Rose, Suzanna., Johnathan Bisson, and Simon Wessely. 2003. A systematic review of single-session psychological interventions ('debriefing') following trauma. *Psychotherapy and Psychosomatics* 72: 176–184.

Shamay-Tsoory, Simone G. 2010. The neural bases for empathy. *The Neuroscientist* 17: 18–24.

Shepard, Roger N. 1990. *Mind sights: Original visual illusions, ambiguities, and other anomalies, with a commentary on the play of mind in perception and art.* New York: W.H. Freeman.

Snow, Charles Percy. 1964. *The two cultures and a second look: An expanded version of the two cultures and the scientific revolution).* Cambridge: Cambridge University Press.

Stoet, Gijsbert, and Geary, David C. 2015. Sex differences in academic achievement are not related to political, economic, or social equality. *Intelligence* 48: 137–151.

Straus, Murray A. 2007. Processes explaining the concealment and distortion of evidence on gender symmetry in partner violence. *European Journal on Criminal Policy and Research* 13: 227–232.

———. 2008. Bucking the tide in family violence research. *Trauma, Violence, & Abuse* 9: 191–213.

———. 2009. Current controversies and prevalence concerning female offenders of intimate partner violence: Why the overwhelming evidence on partner physical violence by women has not been perceived and is often denied. *Journal of Aggression, Maltreatment & Trauma* 18: 552–571.

Sunstein, Cass R. 2009. *Going to extremes: How like minds unite and divide.* Oxford University Press.

Wilson, David S. 2003. *Darwin's cathedral: Evolution, religion, and the nature of society.* University of Chicago Press.

Wilson, Edward O. 1975. *Sociobiology: The new synthesis.* Cambridge, MA: Harvard University Press.

———. 1998. *Consilience: The unity of knowledge.* New York. Knopf.

Wilson, Timothy. 2011. *Redirect: The surprising new science of psychological change.* New York: Penguin.

Wright, Rogers, and Nicholas Cummings, eds. 2005. *Destructive trends in mental health: The well-intentioned path to harm.* New York: Brunner-Routledge.

9 FROM ACTOR TO AGENT TO AUTHOR

HUMAN EVOLUTION AND THE DEVELOPMENT OF PERSONALITY

Dan P. McAdams

In *The Social Conquest of Earth*, Edward O. Wilson (2012) imagines human prehistory as a two-million-year sprint to a unique brand of *eusociality*. Species deemed to be eusocial live in intricately coordinated, multigenerational groups, like beehives and ant colonies. Group members take on specific tasks that contribute to the survival and well-being of the group, and they may engage in altruistic acts that potentially exact an individual cost while benefiting the group. The characteristic *human* manner of eusociality, however, involves the dynamic coordination of highly individuated and cognitively gifted group members, whose implicit goals for self-preservation and self-replication often conflict with the broader aims of the group as a whole. Put simply, human beings are extraordinarily intelligent and agentic, and their social groups are complex, variable, and (often) contentious. From our australopithecine forerunners through *Homo habilis, Homo erectus*, and *Homo sapiens*, then, the presumed course of human evolution has run relentlessly in the direction of increased cognitive power for individual organisms and the formation of increasingly complex and differentiated social groupings (Dunbar 2010).

It is part and parcel of human nature, therefore, to identify closely with groups, for throughout human evolution, individual survival has depended on the survival of the group as a whole and, more important, on one's particular standing within the group. Accordingly, contemporary social psychologists assert that human beings are naturally endowed with a strong *need to belong*—an irresistible desire for attachment to families, clans, teams, tribes, political parties, professional organizations, and all sorts of organized social collectives (Baumeister and Leary 1995). Again and again, experiments in social psychology show that people will develop strong allegiances to virtually any kind of group, even when a group's composition is generated randomly. Groups provide individual members with a social identity, which consists of the individual's specific knowledge regarding how he or she fits into the group and the emotional resonance that such an affiliation provides (Tajfel 1981). Through a wide range of psychological and cultural mechanisms, individual group members develop strong allegiances to their respective groups, believe their groups to be morally superior to other groups, rally to their group's defense during times of threat, and, in general, work hard to advance their group's particular goals and interests (Kesebir 2012).

When considering the reproductive success and inclusive fitness of an individual member of a human group, however, simply being part of the group is not enough. In the case of human eusociality, a group member must find ways to be *accepted* by others in the group while also attaining some modicum of *status*. Throughout human evolution, successful group life for the individual has always depended on his or her ability to *get along* and to *get ahead*—to get along, because social rejection in hominin groups severely compromises the individual's chances for survival; and to get ahead, because attaining some degree of status within the group increases the likelihood of securing one's needed share of limited resources, be those resources food, shelter, or prospective mates (Hogan et al. 1985). Human beings are mandated by human nature to seek social acceptance and social status, to seek to be liked by others, and to achieve a heightened position in a social hierarchy. As the eminent personality psychologist Robert Hogan puts it, "getting along and getting ahead are the two great problems in life that each person must solve" (Hogan et al. 1985, 178). These "fundamental processes" are "rooted in our history as group-living animals," expressed as "archaic, powerful, compulsive tendencies that are closely tied to our chances for survival and reproductive success" (Hogan et al. 1985, 181).

As an individual strives to get along and get ahead in a complex social group, he or she gradually builds up a *reputation* among group members. One member of the group may be seen by others to be especially friendly and helpful. Another may be viewed to be aggressive. Yet another is seen as generally nice but painfully shy and submissive. The human brand of eusociality demands that group members be especially sensitive to consistent variations in social behavior exhibited by other group members, for predicting the behavior of others and knowing things like who can be trusted and whom to avoid are crucial to one's own success in the group, critical to promoting one's own agenda for getting along and getting ahead. Group members are keen to obtain reputational information about other group members, and often keen to share it widely, through gossip, for example (Dunbar 2004). Reputational information distinguishes one group member from another, sketching in rough terms the ways in which each group member is unique, at least as seen from the assiduous gaze of fellow group members.

Which brings us to this chapter's central topic: human *personality*. From an evolutionary standpoint, *personality begins with the different reputations that human beings achieve as they strive to get along and get ahead in social groups.* Natural selection may have conferred an overall design for human nature, but each individual group member is a unique *variation on the design*. Given the peculiarly human form that eusociality has taken throughout the course of evolution, taking note of those variations—monitoring them carefully and sharing information about them through gossip and other means—is a fundamental function the human brain has evolved to perform.

Personality, then, is about the consistent variations in human behavior, thought, and feeling that matter most for getting along and getting ahead in human groups. It is about those variations on the general design for human nature that observant group members have been monitoring ever since *Homo erectus* first walked on the African continent, and those very same variations we have been gossiping about ever since language entered the scene. In what follows, I consider issues and questions raised by evolutionary theory as I describe the development of human personality across the life course. My exposition draws on recent empirical findings in personality, developmental, social, and cognitive psychology, framed within the context of consilience provided by evolutionary theory (Wilson 1998). Reflecting, moreover, a broad and integrative conception of human personality that has gained wide acceptance in recent years (e.g., McAdams 1995; McAdams and Pals 2006; McAdams and Olson 2010), I organize what follows under three general headings: personality from the standpoints of (a) the social actor, (b) the motivated agent, and (c) the autobiographical author.

The Social Actor

Shakespeare was profoundly right when he wrote that all the world's a stage and each of us an actor upon it. For the human brand of eusociality, individual group members perform scripted behaviors on the theatrical stage of social life, entering and exiting ritualized social scenes and enacting scripts and roles in accord with group norms and demands (Goffman 1959; Hogan et al. 1985). Actors strive to get the performance right, for repeatedly botching one's lines or failing to master one's roles can have negative consequences down the road for inclusive fitness. Actors monitor each other's performances, careful to observe and note consistent trends in the expression of behavior, feeling, and desire. Observations are shared through gossip and other forms of social communication and, over time, the actor gets a reputation in the group: there he goes again! Ah, we know her by the unique style she brings to the performance. In human groups, then, the quality of the actor's performances, summarized in social reputation, may be the strongest predictor of his or her facility to get along and to get ahead.

Developmentally speaking, the first inklings of an actor's characteristic performance style may be observed in infancy. Individual differences in *temperament* begin to appear even in the first year of life (Rothbart and Bates 1998). Research in developmental psychology confirms what parents and other audience members have known since the days of *Homo erectus*: Some babies are consistently smiley and cheerful; others express irritability and negative emotional states repeatedly. When the curtain rises, still others seem slow to warm up. Caregivers and other human performers in the baby's life react strongly to these observed differences, rooted as they are in genetic differences among infants, and the repeated reactions of others combine with a wide range of epigenetic and environmental factors to influence the gradual development of basic personality traits across the childhood and adolescent years (Caspi et al. 2005). Through repeated gene-by-environment interactions, early temperament differences gradually morph into the signature personality traits that shape social performance in the adult years.

After half a century of statistical research on the topic, personality psychologists today tend to agree that the basic dimensions of social performance can be reduced to five fundamental categories of dispositional traits, popularly known as the *Big Five* (Goldberg 1993; McCrae and Costa 2008).

The first of these dimensions is *extraversion versus introversion*. Variations on this general factor capture the extent to which a social actor exudes positive emotion, enthusiasm, spontaneity, and social dominance versus the extent to which he or she appears withdrawn, generally quiet, or aloof. Because they go to the heart of human sociality, individual differences in extraversion are among the most readily detected variations in personality. A second dimension is commonly termed *neuroticism versus emotional stability*. A better name, however, is probably "negative affectivity," for this factor captures characteristic differences in the tendency to experience a wide range of negative emotions—from anger and hostility to fear, anxiety, guilt, shame, and sadness. Of course, everybody feels these negative emotions at one time or another in life, but individuals who score high on neuroticism scales tend to experience negative affect repeatedly and with relatively greater intensity compared with individuals lower in neuroticism. In terms of social performance, actors high in neuroticism are known for their consistently negative moods, their emotional volatility and vulnerability, their hypersensitivity to negative life events, and their tendency to make others around them feel almost as anxious and depressed as they often feel. By contrast, those low in neuroticism generally appear to be emotionally stable, calm and collected, stolid in the face of adversity, and resilient. Extraversion and neuroticism pair up to account for a great deal of variance in the emotional quality of actors' performances, with extraversion capturing positive emotion in the company of others and neuroticism capturing a more negativistic and anxious style of social performance.

Although extraversion and neuroticism tap the emotional wellsprings of human social interaction, the third and fourth dimensions of the Big Five are more about everyday behavior in social roles. *Agreeableness* (*versus disagreeableness*) speaks to the extent to which a social actor is warm, kind, caring, and altruistic in his or her dealings with others versus the extent to which he or she is cold-hearted, mean-spirited, and cruel. *Conscientiousness* (*versus nonconscientiousness*) subsumes a broad assortment of variations in the characteristic performance of social roles and instrumental tasks. People high in conscientiousness are hard-working, focused, well-organized, and highly disciplined; those scoring low on this dimension are the opposite—tending toward laziness and entropy. Not surprisingly, individual differences in conscientiousness are strong predictors of success in school and in the workplace (Barrick and Mount 1991). Agreeableness and conscientiousness are markers of social maturity. Research suggests that as people develop through adolescence and into early and middle adulthood, their scores on conscientiousness and agreeableness often rise, reflecting their increasingly generative involvement in such adult social roles as spouse, parent, and citizen (Roberts et al. 2006; Lodi-Smith and Roberts 2007).

The fifth cluster of personality traits typically goes by the name of *openness to experience*. Variation on this dimension runs from being especially open-minded, intellectually curious, nonconventional, and aesthetically oriented on one end to dogmatic, practical, concrete, and conventional on the other. Openness taps into more cognitive and attitudinal dimensions of social performance. It is also modestly, but typically significantly, correlated with objective measures of general intelligence, with people high in openness sometimes enjoying higher levels of IQ. Although precursors to extraversion, neuroticism, conscientiousness, and agreeableness can be observed in temperament dimensions of young children, openness is harder to track in a social developmental sense. Nonetheless, variation on this dimension is readily observed in everyday social interaction, especially when human beings get together to solve problems or to learn new skills.

Evolutionary thinking about individual differences in dispositional personality traits has tended to follow two different lines. One argument contends that variation on such traits as extraversion and agreeableness is of no evolutionary significance whatsoever, but merely reflects random noise in human nature (Tooby and Cosmides 1990). A second line of thought suggests that such variation is a prime example of *frequency-dependent selection* (Gangestad and Simpson 1990; Cochran and Harpending 2009). In frequency-dependent selection, the fitness of a phenotype is dependent on its frequency to other phenotypes in a given population. There are variations that have positive effects when rare, but with advantages that decrease as they become more common, eventually becoming negative, as modeled, for example, in computer simulations of what is called the hawk-and-dove game. When aggressive hawks are rare, they easily defeat peaceful doves and have greater fitness. As they become more common, however, hawks confront other hawks more often and end up in costly fights that decrease their fitness. At some frequency in the population, the fitness of hawks and doves is the same, leading to a kind of balance. Similarly, it has been argued that human sociopaths—individuals who are well designed to be cheaters, such as con artists—can prosper when they are rare but suffer fitness loss as they become more common and other actors become more aware of them. A parallel logic may be applied to the Big Five dispositional traits. Low levels of, say, conscientiousness or agreeableness may go unpunished in fitness terms as long as there are many hard-working and empathic individuals in the group. But too many nonconscientious and disagreeable social actors—lazy and mean-spirited as they are—decreases the fitness payoff for these phenotypes. And so trait variation may prevail, with each spot on the bell-shaped continuum for the distribution of a given trait representing a potentially adaptive dispositional niche.

Regardless of whether evolutionary conceptions can help to explain the actual variation that exists in levels of dispositional personality traits, it seems clear that human beings evolved to notice and respond to these differences, and to accord them special attention (Buss 1996; McAdams and Pals 2006). The original derivation of the Big Five taxonomy came from studies of lexicons, under the assumption that important individual differences in personality should be expressed in the common words used in human language (Goldberg 1981). Studies of many different language groups show that the Big Five factor structure, or something very similar to it, appears in lexicons the world over (McCrae and Costa 2008). Although each social actor may express a unique dispositional profile, therefore, *the tendency to perceive actors' profiles in terms of five basic dimensions of social performance may itself be a feature of human nature.* Indeed, the five basic dimensions provide the kind of first-cut information that observers are likely to want to know about the actors they observe and with whom they are likely to interact (Goldberg 1981; McAdams 1992). For extraversion: How socially dominant is X? Will X dominate me? To what extent might I gain social resources by allying myself with X? For neuroticism: How emotionally stable is X? When things get tough, can I count on X to be strong? For agreeableness: Can I trust X? Will X help me? Will X care for me? For conscientiousness: Can I depend on X? Will X work hard with me? And for openness to experience: How curious and flexible is X? Will X be a good resource for me when I confront new problems?

To get along and get ahead in social groups, human beings need to find answers to questions such as these. Furthermore, as actors on the social stage of life, human beings forge reputations in their groups that, among other things, spell out where the actor is likely to stand on each of the five basic trait dimensions. As social actors, we know each other first, and in many cases best, by the dispositional traits we observe and express.

The Motivated Agent

Actors have secrets that observers can never fully know. From the vantage point of the theater seats, the audience can directly observe what the characters do on stage, but the audience can only *infer* their intentions, desires, and goals. Even if a character proclaims what he wants in a long soliloquy or sudden exclamatory burst, there is always a chance he is lying or, more interesting, he does not consciously know what he really wants, does not understand his true motivations. Art imitates human life in this sense, for discerning the intentions of others is a quintessentially human challenge, a cardinal prerequisite for getting along and getting ahead in human social groups. The task of reading others' minds requires a tremendous amount of cognitive processing capacity and, thus, may be one big reason that human beings have such big brains:

> Over the last forty years many have proposed that the pressure to understand others has been a major driving force in the growth of higher, especially primate, intelligence. We need to infer others' predispositions and their likely intentions and actions, which may in turn be based in part on *their* attempts to guess *ours*. If this hypothesis is correct, higher intelligence emerged primarily as social intelligence, through a cognitive arms race to understand conspecifics and to reveal to or conceal from others our own beliefs, desires, and intentions. (Boyd 2009, 141)

The dispositional traits that make up the first layer of human personality address the social actor's overall performance style, but they say little, if anything, about what the person who displays those traits really wants in life—that is, they say little about the vicissitudes of human *agency*. Knowing that a person is high in extraversion, for example, tells you little about the person's goals, values, fears, and intentions. An extravert might strive to become a great basketball player, a dutiful son, a journalist, an artist, a devout Catholic, a body-builder, a loving father, a person who cares little about money or fame, or president of the United States. An introvert might strive for the very same things. Layered over dispositional traits, then, are the characteristic goals, values, plans, and intentions that speak to a person's status as a motivated agent. If the dispositional traits encompassed within the Big Five taxonomy form the base of psychological individuality, the particulars of a person's agency make up personality's second layer.

Agency refers to matters of choice, intention, desire, and motivation in human affairs (Bandura 1989). To be an agent is to act in accord with one's volition or will, to pursue self-determined goals, whether they be goals to approach desirable end states or to avoid undesirable ones. In a rudimentary sense, even human infants behave in a goal-directed fashion (Bowlby 1969). But human beings do not become consciously aware of their ontological status as motivated agents until they are three or four years of age. And they do not typically organize their daily activities, and consistently understand who they are, in terms of goals, plans, and values until they are well into the elementary school years.

By age one year, human infants express a basic appreciation for agency. Research has shown, for example, that nine-month-old infants can distinguish between intentional and accidental behavior on the part of caregivers (Behne et al. 2005) and prefer to imitate intentional behavior over random acts (Woodward 2009). By age four years, most children have internalized what developmental researchers call a *theory of mind* (Wellman, 1993). Before they enter kindergarten, children have developed an implicit theory about how minds (their own and others) operate and why people do what they do. Their simple folk–psychological theory says that people have beliefs and desires in their minds that ultimately motivate their behavior. In other words, children now expressly conceive of people as motivated agents: People do things because they want to do them (desire) and on the basis of internal beliefs they have about things.

Equipped now with theory of mind, children may promiscuously project agency onto many other beings and things in their environment, including inanimate objects, imaginary companions, and culturally mediated concepts such as God (Kelemen 2004). In a reflexive manner, furthermore, children are now able to appreciate organized agency in themselves. Applying theory of mind to the self, the child is now able to see how his or her own desires, goals, and wants motivate his or her own behavior. By age eight or nine years, children begin to define and evaluate themselves in terms of their personal goals, projects, and values (McAdams and Cox 2010). Their self-esteem rises and falls as a function of how well they believe they are doing with regard to the goals they value most (Harter 2006). By late childhood, then, personality thickens to accommodate a second layer of personal attributes that speak directly to the person's status as a motivated agent in society.

It is no accident that the basic human motives and goals identified by personality researchers today—those same motives and goals that begin to emerge as key features of personality during the late childhood years—track closely the dualistic challenge for which evolution has designed our eusocial species: getting along and getting ahead in social groups. Going back even to Freud's (1961/1930) famous dichotomy of Eros and aggression, researchers have distinguished repeatedly between two classes of basic human motivations: those designed to promote communion, love, intimacy, affiliation, group bonding, and interdependence on the one hand; and those aimed to promote individual control, power, status, achievement, self-expansion, and independence on the other (Bakan 1966; Wiggins and Trapnell 1996).

Whether considering the contrast between, say, intimacy motivation and power motivation (McAdams 1985), relationship goals and autonomy goals (Deci and Ryan 1991), or interdependent and independent value systems (Markus and Kitayama 1991), the dynamic between group affiliation and individual expression permeates psychological discourse, from the recurrent themes expressed in social–scientific research to psychotherapy talk, Hollywood movies, and popular fiction. Finding a way to reconcile potentially discordant concerns for the group and the self appears to be a human universal, metaphorically paralleling what Wilson (2012) characterizes as the evolutionary dynamic between group selection and individual selection. Those individual human beings who regularly accomplish reconciliation between the two typically enjoy higher levels of psychological well-being and maturity. For example, empirical studies have shown that adults who are deemed to be exemplars of psychological health and moral commitment tend to achieve a dynamic balance between other-oriented and self-oriented motivations. Research examining the motives, goals, values, and self-descriptions of men and women who are widely admired for their positive contributions to society shows that human paragons of well-being and virtue repeatedly find ways to use their strong needs for power and achievement for the good of others (McAdams et al. 1986; Colby and Damon 1992; Frimer et al. 2011, 2012). By getting ahead, they also manage to get along—and manage, at the same time, to help other members of their constituent groups in their own respective efforts to get ahead and to get along.

The Autobiographical Author

In the development of human personality, *to be an agent one must first be an actor.* From the beginning, human beings act in a social context; later, they figure out *why* they are acting (or why they think they are acting) and develop plans and goals to structure their acting in the future. Eventually, human beings become authors, too, narrating sequences of action and agency to themselves and to others. People tell stories about what they did and why they did it, and they articulate broader narratives about how they believe their lives have unfolded over time, and what their lives—past, present, and future—may mean.

To be an author, one must first be an agent, for the very nature of human storytelling involves narrating how a motivated character seeks to enact intentions over time (Bruner 1986). "Characters are individual agents with goals," writes Carroll (2008, 118). "Novelists and playwrights are individual persons who construct intentional meanings about those characters . . . [therefore, it] is not possible to speak of depicted narrative events without at least tacitly identifying agents and goals" (Carroll 2008, 118). Whether telling the story about why, in spring 1989, she decided to get married, or writing a novel about love and war in 16th-century England, the storyteller provides an account of motivated agents who make decisions, seek goals, articulate dreams and fears, and struggle to realize their aspirations in a world in which other motivated agents aim to realize their own, often competing, agendas, played out across social settings and over time. Authorship, then, presupposes agency, which presupposes acting.

The developmental roots of personal authorship go back as far as the second and third years of human life, with the development of language and the emergence of autobiographical memory (Howe and Courage 1997; Fivush 2011). Before the age of two years, children do not encode and store stable memories about personal events, although they recognize familiar stimuli encountered in the past. With the advent of autobiographical memory, however, children begin to remember events in their lives as having happened to *them*; they begin to claim explicit ownership of personal experiences, recalling that I, *myself*, witnessed something happen in *my* life, an experience that may now be deemed *mine*. Children collect little memories about their own personal experiences, and they eventually begin to tell stories about those memories to others, encouraged by parents, family members, teachers, and nearly everybody else who engages in conversations with children. As they have more and more conversations, children learn how to tell better and better (more convincing, coherent, entertaining) stories about the past.

With the consolidation of theory of mind in the third and fourth years of life, children are able to confer upon the characters in their personal accounts the clear sense of agency that good stories typically express. By the time they are five or six years of age, most children are able to tell stories about themselves (and about others) that conform to what cognitive scientists call a *narrative grammar*. They now know, for example, that stories are typically set in a particular time and place, begin with a motivated protagonist who sets off to accomplish a goal, become interesting when the protagonist encounters obstacles to goal achievement, become complicated further when the protagonist responds to the obstacles, and are eventually resolved when the goal is either achieved or finally thwarted. Stories are supposed to engender suspense, and they are supposed to end with a release of tension and a sense of closure. When narrative accounts, then, do *not* conform to these common expectations about stories, young children may find them confusing, and may later misremember the accounts as having a more canonical form than they originally had (Mandler 1984).

It is one thing to tell a story about a personal experience; it is a much bigger thing to tell a story about a full life. To tell a full life story, a person must be able to engage in what Habermas and Bluck (2000) call *autobiographical reasoning*. In autobiographical reasoning, a narrator articulates organizing themes, derives causal explanations, and draws convincing inferences about the self from and through the telling of an extended narrative about life. It is not until adolescence that most people show the forms of autobiographical reasoning that are required to construct and internalize full life stories (Habermas and de Silveira 2008; Pasupathi and Wainryb 2010). Moreover, adolescence is a time in the human life course in which people typically encounter such identity questions as: Who am I? How did I come to be? Where is my life going? What does my life really mean? The development of a broad narrative understanding for one's life serves to address identity questions such as these.

In adolescence, therefore, many people begin to articulate and internalize a story about the self that explains how the person came to be and where his or her life may be going in the future. Psychologists call this story a *narrative identity* (McAdams 1985; Singer 2004; McLean et al. 2007). A narrative identity, then, is an evolving and internalized story of the self that reconstructs the past and imagines the future in such a way as to provide life with some degree of unity and purpose. In narrative identity, a person authors an integrative story for life. The story tells who I am, who I was, and who I may be in the future. The story helps to explain why the social actor does what it does, why the motivated agent wants what it wants, and what the author's life means in the context of time, personal experience, society, and culture. In constructing a narrative identity, the autobiographical author draws from a menu of favored themes, images, motifs, plots, and characters that prevail in his or her given society and culture (McAdams 2006; Hammack 2008). The life stories that authors tell reflect both their own personal experiences, as social actors and motivated agents, and the expectations and norms for life storytelling that prevail within the psycholiterary contexts embedded in culture, class, religion, ethnicity, gender, and the unique historical moment within which an individual life is lived (Rosenwald and Ochberg 1992).

Because people tell stories in all human societies, scientists and scholars have speculated broadly about the possible adaptive functions of storytelling in the context of human evolution. For example, Pinker (1997) has argued that humans tell stories mainly for the sake of pleasure. From Pinker's perspective, the human proclivity to tell and enjoy stories would seem to have little specific evolutionary significance. In contrast, Boyd (2009) contends that storytelling may stimulate cognitive development and creativity. Mar and Oatley (2008) conceive of storytelling as the abstraction and simulation of social experience. Within the evolutionary context of human eusociality, they argue, telling stories is a powerful means whereby group members can represent and convey their actual experiences of group life, to themselves and to each other, to understand those experiences better and mine them for insights and lessons that could be applied to future social behavior.

Consistent with the views of Boyd (2009) and Mar and Oatley (2008), stories can also be used as testing grounds for hypothetical scenarios: What might happen if I formed an alliance with this person or that person? How would life change if I were to take on this project or that project? By entertaining stories about what *might* happen, narrators can explore contingencies and contexts while safely projecting themselves into alternative futures. In so doing, they can imagine in their own narrating minds how a particular line of action and agency might play out and what its implications might be for their own efforts to get along and get ahead. More important, storytellers can obtain useful feedback from sympathetic audiences, with whom stories are shared and from whom reactions are monitored. Telling stories is also a favored practice for passing on values and virtues to others (MacIntyre 1981). Contemporary research in the social and behavioral sciences underscores the efficacy of storytelling in the moral socialization of children (Narvaez and Lapsley 2009), the rehabilitation of criminals (Maruna 2001), and healing work of psychotherapy and counseling (Lieblich et al. 2004). Even under the aegis of cultural modernity, the ancient art of storytelling appears to be an invaluable resource for enhancing human adaptation to social life. It is, therefore, a rather small inferential leap to suppose that stories might have promoted successful adaptation to group life among our evolutionary ancestors.

Although narrative identity is surely an outgrowth of the universal human tendency to tell stories about personal experiences, and although such a tendency may have conferred adaptive advantages over the course of human evolution, it does not necessarily follow that narrative identity itself is an evolutionary adaptation. Indeed, research and theory on narrative identity has tended to emphasize the particular ways in which certain cultural arrangements call for the construction of integrative life stories, whereas other cultural arrangements do not (Gergen 1991; McAdams 1996, 2006). More specifically, it would appear that the need to articulate a self-defining life narrative—an integrative and unique personal story that reconstructs the past and imagines the future to provide an individual life with unity, purpose, and identity—may be peculiar to, or at least especially salient within, modern and postmodern societies, such as contemporary North America and Europe, Japan, Australia, and parts of the developing world.

As societies advance in science and technology, as they develop open markets and democratic political institutions, and as they generate manifold life possibilities that defy a common consensus regarding how a good human life should be lived, the adolescent and adult members of those societies are increasingly challenged to construct narrative identities to provide individual lives with purpose, meaning, and order (Giddens 1991; McAdams 1996). Narrative identity would appear to be an outgrowth of cultural conditions wherein people are strongly urged (a) to think about who they really are and what their lives really mean, (b) to find or develop their own unique niche amid a panoply of instrumental and expressive life options, and (c) to construe the human life course itself as a psychological journey of self-exploration, aimed toward personal fulfillment. Storytelling may have been a universally prized human activity since the evolution of human language—and maybe even before, to the extent stories were originally expressed in dance and ritual. However, constructing an integrative narrative for one's life as a whole was probably a psychological challenge that was not so strongly emphasized under premodern feudal conditions, in traditional agrarian and pastoral contexts, and among our hunting and foraging forebears. Although its roots, therefore, may lie in the evolved capacities of a cognitively advanced eusocial species, narrative identity assumes an especially salient position under the aegis of cultural modernity, reflecting the unique challenges that modern humans face in formulating a meaningful life in an ambiguous and multifaceted cultural world.

Conclusion: Personality and the Circles of Sociality

Current scientific thinking on the development of personality suggests that human beings are first and foremost social actors, whose dispositional performance tendencies shape and are conveyed through their social reputations. During late childhood, human beings become motivated agents, too, as self-chosen goals and values layer over basic dispositional traits. Although each person develops his or her own unique motivational agenda, personal goals often track the two fundamental challenges of human eusociality: getting along and getting ahead in the context of the group. Under the conditions of cultural modernity, a third layer of personality begins to emerge in the adolescent and young adult years. As autobiographical authors, adolescents and adults construct integrative life stories—narrative identities—designed to explain who they are, how they came to be, and where they believe their lives are going in the future. These internalized and integrative stories of the self are especially designed to meet the daunting identity challenges of modern life. The powers to construe one's life in full as a synthetic narrative, however, derive from basic evolutionary adaptations regarding language development, autobiographical memory, and the cognitive skills required for abstracting and simulating social experience.

Personality begins with the dispositional traits that define the actor's social reputation for and in the group. Because social observation invariably results in the attribution of traits to actors, trait reputations are formulated readily and communicated easily in groups, through gossip and other forms of information sharing. Therefore, one does not need to have an especially intimate understanding of an individual group member to know something useful about that individual's respective traits. As the knowledge that one has about another group member, or even about the self, deepens, however, one encounters the stuff of layer 2 in personality—the goals, values, plans, and envisioned projects that define the motivated agent's personal agenda for getting along and getting ahead in the group (McAdams 1995). Although a relatively large number of group members may know a person's traits, it is a smaller number of group members who know the person well enough to understand his or her goals, values, and motives. Fewer still will be privy to the stories a person owns and makes about his or her own life, the deeply emotional remembrances of things past, the private scars and the personal triumphs, the roads not taken, the dreams unfulfilled, the beautiful moments of triumph and insight captured in episodic flashbacks, and elaborated disquisitions about the self. Long before citizens of modern societies struggled to formulate their narrative identities, *Homo sapiens* formed and told stories about their personal experiences, presumably sharing the most private and personal stories with but a small group of intimates.

The three layers of personality, therefore, roughly correspond to three levels of sociality that Dunbar and Sutcliffe (2012) suggest characterized group life among our evolutionary ancestors. The broadest circle of human sociality in hunting and foraging societies was the *activity network*. This broad circle of associates might consist of all those individuals with whom a given person has a personal relationship of some sort, no matter how rudimentary—typically no more than about 150 individuals, Dunbar and Sutcliffe (2012) maintain. In terms of the three layers of personality described in this chapter, these are the people who have garnered reliable information about the social actor's reputation. They know the person by his or her traits.

Closer in is what Dunbar and Sutcliffe (2012) describe as the *affinity group*—the approximately 50 people within the activity network with whom the person regularly engages in joint goal pursuits—fellow *agents* who share a motivational agenda, working together, perhaps, on common projects, plans, and programs. In addition to their knowledge of the person as a social actor, this more exclusive group relates to the person in question in terms of joint motivational agendas. They know the person's goals, plans, and values, in addition to the person's dispositional traits. They need to know the person in this deeper and more intensive way, because their prospects of getting along and getting ahead in group life depend, in part, on their working successfully with the person in question, coordinating multiple motivational agendas.

In the smallest and closest circle reside those members of a person's *sympathy group* (Dunbar and Sutcliffe 2012). These are the 5 to 10 people with whom the person is closest emotionally and with whom the person most likely *shares a history*. These closest friends and family members spend the most time with the person and provide emotional support for him or her during times of need. They are the most likely members of the group to care for the person, to love him or her to the point of readily and repeatedly sacrificing their own well-being for the good of the person. The members of the inner circle know the person by virtue of his or her traits, goals, and narratives. They know the most important events in the person's autobiographical past, and they have some sense of how he or she imagines his or her future, as well. They know the person best because they have experienced some of the same events he or she has experienced, perhaps by virtue of having grown up with the person or having cared for him or her when he or she was young. They know the person best because they know how he or she typically thinks and feels, what the person really wants and fears, and how he or she understands his or her own life and the world. They know the person best because they know his or her stories.

References

Bakan, David. 1966. *The duality of human existence*. Boston: Beacon Press.

Bandura, Albert. 1989. Human agency in social-cognitive theory. *American Psychologist* 44: 1175–1184.

Barrick, Murray R., and Michael K. Mount. 1991. The Big Five personality dimensions and job performance: A meta-analysis. *Journal of Personnel Psychology* 44: 1–26.

Baumeister, Roy F., and Mark R. Leary. 1995. The need to belong: Desire for interpersonal attachment as a fundamental human motivation. *Psychological Bulletin* 117: 497–529.

Behne, Tanya, Malinda Carpenter, Joseph Call, and Michael Tomasello. 2005. Unwilling versus unable: Infants' understanding of intentional action. *Developmental Psychology* 41: 328–337.

Bowlby, John. 1969. *Attachment.* New York: Basic Books.

Boyd, Brian. 2009. *On the origin of stories: Evolution, cognition, and fiction.* Cambridge, MA: Harvard University Press.

Bruner, Jerome. 1986. *Actual minds, possible worlds.* Cambridge, MA: Harvard University Press.

Buss, David M. 1996. Social adaptation and the five factors of personality. In Jerry S. Wiggins, ed., *The five-factor model of personality: Theoretical perspectives.* New York: Guilford Press.

Carroll, Joseph. 2008. An evolutionary paradigm for literary study. *Style* 2: 103–135.

Caspi, Avshalom, Brent W. Roberts, and Rebecca Shiner. 2005. Personality development: Stability and change. In Susan T. Fiske and Daniel Schacter, eds., *Annual review of psychology*, 453–484. Palo Alto, CA: Annual Reviews.

Cochran, Gregory, and Henry Harpending. 2009. *The 10,000 year explosion: How civilization accelerated human evolution.* New York: Basic Books.

Colby, Anne, and William Damon. 1992. *Some do care: Contemporary lives of moral commitment.* New York: Free Press.

Deci, Edward L., and Richard M. Ryan. 1991. A motivational approach to self: Integration in personality. In Richard Dienstbier and Richard M. Ryan, eds., *Nebraska symposium on motivation*, 237–288. Lincoln: University of Nebraska Press.

Dunbar, Robin 2004. Gossip in evolutionary perspective. *Review of General Psychology* 8: 100–110.

———. 2010. *How many friends does one person need? Dunbar's number and other evolutionary quirks.* Cambridge, MA: Harvard University Press.

Dunbar, Robin, and Alistair Sutcliffe. 2012. Social complexity and intelligence. In Jennifer Vonk and Todd K. Shackelford, eds., *The Oxford handbook of comparative evolutionary psychology*, 102–117. New York: Oxford University Press.

Fivush, Robyn. 2011. The development of autobiographical memory. In Susan T. Fiske, Daniel L. Schacter, and Shelley E. Taylor, eds., *Annual review of psychology* 559–582. Palo Alto, CA: Annual Reviews.

Freud, Sigmund. 1961 [1930]. *Civilization and its discontents.* Translated and edited by James Strachey. London: Hogarth.

Frimer, Jeremy A., Lawrence J. Walker, William L. Dunlop, Brenda H. Lee, and Amanda Riches. 2011. The integration of agency and communion in moral personality: Evidence of enlightened self-interest. *Journal of Personality and Social Psychology* 101: 149–163.

Frimer, Jeremy A., Lawrence J. Walker, Brenda H. Lee, Amanda Riches, and William L. Dunlop. 2012. Hierarchical integration of agency and communion: A study of influential moral figures. *Journal of Personality* 80: 1117–1145.

Gangestad, Steven W., and Jeffrey A. Simpson. 1990. Toward an evolutionary history of female sociosexual variation. *Journal of Personality* 58: 69–96.

Gergen, Kenneth J. 1991. *The saturated self: Dilemmas of identity in contemporary life.* New York: Basic Books.

Giddens, Anthony. 1991. *Modernity and self-identity: Self and society in the late modern age.* Stanford, CA: Stanford University Press.

Goffman, Erving. 1959. *The presentation of self in everyday life.* Garden City, NY: Doubleday.

Goldberg, Lewis R. 1981. Language and individual differences: The search for universals in personality lexicons. *Review of personality and social psychology* 2: 141–165.

———. 1993. The structure of phenotypic personality traits. *American Psychologist* 48: 26–34.

Habermas, Tilman, and Susan Bluck. 2000. Getting a life: The emergence of the life story in adolescence. *Psychological Bulletin* 126: 748–769.

Habermas, Tilman, and Cybele de Silveira. 2008. The development of global coherence in life narratives across adolescence: Temporal, causal, and thematic aspects. *Developmental Psychology* 44: 707–721.

Hammack, Philip L. 2008. Narrative and the cultural psychology of identity. *Personality and Social Psychology Review* 12: 222–247.

Harter, Susan. 2006. The self. In Nancy Eisenberg, William Damon, and Richard Lerner, eds., *Handbook of child psychology*, vol. 3, 505–570. New York: Wiley.

Hogan, Robert, Warren H. Jones, and Jonathan Cheek. 1985. Socioanalytic theory: An alternative to armadillo psychology. In Barry R. Schlenker, ed., *The self and social life*, 175–198. New York: McGraw-Hill.

Howe, Mark L., and Mary L. Courage. 1997. The emergence and early development of autobiographical memory. *Psychological Review* 104: 499–523.

Kelemen, Deborah. 2004. Are children "intuitive theists?" *Psychological Science* 15: 295–301.

Kesebir, Selin. 2012. The superorganism account of human sociality: How and when human groups are like beehives. *Personality and Social Psychology Review* 16: 233–261.

Lieblich, Amia, Dan P. McAdams, and Ruthellen Josselson, eds. 2004. *Healing plots: The narrative bases of psychotherapy.* Washington, DC: American Psychological Association.

Lodi-Smith, Jennifer, and Brent W. Roberts. 2007. Social investment and personality: A meta-analysis of the relationship of personality traits to investment in work, family, religion, and volunteerism. *Personality and Social Psychology Review* 11: 68–86.

MacIntyre, Alasdair. 1981. *After virtue.* Notre Dame, IN: University of Notre Dame Press.

Mandler, Jeanne M. 1984. *Stories, scripts, and scenes: Aspects of schema theory.* Hillsdale, NJ: Erlbaum.

Mar, Raymond A., and Keith Oatley. 2008. The function of fiction is the abstraction and simulation of social experience. *Perspectives on Psychological Science* 3: 173–192.

Markus, Hazel, and Shinobu Kitayama. 1991. Culture and the self: Implications for cognition, emotion, and motivation. *Psychological Review* 98: 224–253.

Maruna, Shadd. 2001. *Making good: How ex-convicts reform and rebuild their lives.* Washington, DC: APA Books.

McAdams, Dan P. 1985. *Power, intimacy, and the life story: Personological inquiries into identity.* Homewood, IL: Dorsey Press.

———. 1992. The five-factor model in personality: A critical appraisal. *Journal of Personality* 60: 329–361.

———. 1995. What do we know when we know a person? *Journal of Personality* 63: 365–396.

———. 1996. Personality, modernity, and the storied self: A contemporary framework for studying persons. *Psychological Inquiry* 7: 295–321.

———. 2006. *The redemptive self: Stories Americans live by.* New York: Oxford University Press.

McAdams, Dan P., and Keith S. Cox. 2010. Self and identity across the life span. In Richard Lerner, Alexandra Freund, and Michael Lamb, eds., *Handbook of life span development,* vol. 2, 158–207. New York: Wiley.

McAdams, Dan P., Karin Ruetzel, and Jeanne M. Foley. 1986. Complexity and generativity at midlife: Relations among social motives, ego development, and adults' plans for the future. *Journal of Personality and Social Psychology* 50: 800–807.

McAdams, Dan P., and Bradley Olson. 2010. Personality development: Continuity and change over the life course. In Susan Fiske, Daniel Schacter, and Robert Sternberg, eds., *Annual Review of psychology,* 517–542. Palo Alto, CA: Annual Reviews.

McAdams, Dan P., and Jennifer L. Pals. 2006. A new Big Five: Fundamental principles for an integrative science of personality. *American Psychologist* 61: 204–217.

McCrae, Robert R., and Paul T. Costa, Jr. 2008. The five-factor theory of personality. In Oliver P. John, Richard W. Robins, and Lawrence A. Pervin, eds., *Handbook of personality: Theory and research,* 3rd ed., 159–181. New York: Guilford.

McLean, Kate C., Monisha Pasupathi, and Jennifer L. Pals. 2007. Selves creating stories creating selves: A process model of self-development. *Review of Personality and Social Psychology* 11: 262–278.

Narvaez, Darcia, and Daniel K. Lapsley, eds. 2009. *Personality, identity, and character: Explorations in moral psychology.* New York: Cambridge University Press.

Pasupathi, Monisha, and Carolyn Wainryb. 2010. On telling the whole story: Facts and interpretations in autobiographical memory narratives in from childhood through midadolescence. *Developmental Psychology* 46: 735–746.

Pinker, Steven. 1997. *How the mind works.* New York: Norton.

Roberts, Brent W., Kate E. Walton, and Wolfgang Viechtbauer. 2006. Patterns of mean-level change in personality traits across the life course: A meta-analysis of longitudinal studies. *Psychological Bulletin* 132: 1–25.

Rosenwald, George C., and Richard L. Ochberg, eds. 1992. *Storied lives: The cultural politics of self-understanding.* New Haven, CT: Yale University Press.

Rothbart, Mary K., and James E. Bates. 1998. Temperament. In Nancy Eisenberg, ed., *Handbook of child psychology*, vol. 3, 5th ed., 105–176. New York: Wiley.

Singer, Jefferson A. 2004. Narrative identity and meaning making across the adult lifespan: An introduction. *Journal of Personality* 72: 437–459.

Tajfel, Henri. 1981. *Human groups and social categories.* Cambridge, UK: Cambridge University Press.

Tooby, John, and Leda Cosmides. 1990. On the universality of human nature and the uniqueness of the individual: The role of genetics and adaptation. *Journal of Personality* 58: 17–67.

Wellman, Henry M. 1993. Early understanding of mind: The normal case. In Simon Baron-Cohen, Helen Tager-Flusbert, and Donald J. Cohen, eds., *Understanding other minds: Perspectives from autism*, 10–39. New York: Oxford University Press.

Wiggins, Jerry S., and Paul D. Trapnell. 1996. A dyadic-interactional perspective on the five-factor model. In Jerry S. Wiggins, ed., *The five-factor model of personality: Theoretical perspectives*, 89–162. New York: Guilford Press.

Wilson, Edward O. 1998. *Consilience: The unity of knowledge.* New York: Knopf.

———. 2012. *The social conquest of earth.* New York: Norton.

Woodward, Amanda L. 2009. Infants' grasp of others' intentions. *Psychological Science* 18: 53–57.

V A BIOCULTURAL PERSPECTIVE ON LITERATURE

10 WHAT DO ROMANCE NOVELS, PRO WRESTLING, AND MACK BOLAN HAVE IN COMMON?

CONSILIENCE AND THE POP CULTURE OF STORYTELLING

Catherine Salmon

Involvement in fictional, imagined worlds—oral and written stories, drama, film, opera, comic books, jokes, and even ballet—appears to be a human universal (Tooby and Cosmides 2001; Gottschall 2012). What does an evolutionary perspective on the mind entail for our understanding of fiction? Evolutionary psychology offers explanations for how evolution, via natural and sexual selection, has shaped human bodies, minds, and behavior, and how culture has emerged out of our evolved nature. Over evolutionary timescales, humans have faced three main types of problems: problems of survival, problems of mating and parenting, and problems of social living. Whether it's movie plot lines (Garcia et al. 2012), pornography (Salmon and Symons 2001), or a comedian's routine (Kuhle 2012), pop culture focuses on the ways in which people deal (successfully or not) with these same problems in the modern world. For example, horror films (typically viewed through a Freudian psychoanalytic lens [see Clasen 2012, and Chapter 11, this volume) share a cross-cultural appeal. All horror films stimulate fear—an emotion that has evolved through natural selection to help individuals survive and reproduce. Experiencing fear without the risk of real danger is enjoyable for many people. The desire to experience such things secondhand may function to help us better read the minds of others, become more attuned to cues of danger, and become more alert to possible solutions to such dangers (Grodal 2009). Stories allow us to gather a collection of involving problems or conflicts on which we can later draw to solve problems of our own. We observe the problems, the solutions attempted, and their outcomes (Tooby and Cosmides 2001; Gottschall 2012).

Some of the most popular genres of pop culture involve storytelling, such as romance, pro wrestling, and action–adventure novels. Although such genres may be evolutionarily novel, their content is not. Their stories focus on the problems humans have always faced: ones of survival, competition, mating, and social living. Their essential storylines draw in their readers and watchers because they tap in to the evolved psychological mechanisms we have for managing these problems in our own lives. A consilient approach to the study of pop culture can explain why certain problems predominate in stories, why they arouse our emotions, and why there are sex differences in the appeal of certain genres. An adaptationist approach can provide a comprehensive interpretation and understanding of these products of our evolved mind.

Nights (and Knights) in White Satin

Traditional Approaches

The romance novel genre has seldom been taken seriously. Much of the analysis of romance novels in the past has adopted a social constructivist perspective or has been inspired by a feminist ideology. Camp (1997) suggests that romance appeals to women because they teach about nurturing and achieving success or rewards, whereas feminist scholars such as Brownmiller (1976) claim that male conditioning is responsible for women's enjoyment of romance. Some claim they are a venue for rebellion again male authority (Modelski 1980). Other critics have focused on the impact the portrayal of women in romances has on readers. They have suggested that romances have a negative impact, encouraging stereotypical traditional gender roles (Mussell 1984). Other researchers argue that romance is about female empowerment because it depicts women seducing and transforming men (Radway 1984). These explanations add relatively little to our understanding of why romance has been such a popular genre for women over time and in so many different cultures, as seen in the widespread distribution and translation of romances today. A deeper level of explanation appears in Helen Hazen's (1983) underappreciated book *Endless Rapture.* Hazen argues that all women's fiction explores the same basic themes: the problems inherent in finding, acquiring, and retaining a suitable mate.

The Adaptationist View

If one purpose of fiction is the organization of our thinking about ancestral problems, it should not be surprising that a vastly popular type of storytelling would involve the saga of mate choice—in particular, female mate choice. For ancestral women, the greater minimal investment required by women (Trivers 1972) in an offspring means she must be highly selective in her choice of a long-term mate. She must be sure the male who will be contributing his sperm is also willing and capable of contributing the resources required to raise successfully any children produced. As a result, female mating adaptations are sensitive to cues that a male is capable of providing economic resources or has good prospects for doing so in the future (Kenrick et al. 1990; Buss et al. 2001). They also exhibit preferences for men of high status (Li 2007), which is typically associated with access to resources, as well as men who are dependable (Buss et al. 1990) and demonstrate a willingness to invest in children (La Cerra 1994; Roney et al. 2006). Physical height and strength are also seen as desirable (Hughes and Gallup 2003; Courtiol et al. 2010), as are indicators of genetic quality such as facial symmetry and masculinity (Rhodes 2006; Thornhill and Gangestad 2008). In general, romance is focused on the domain of long-term mating.

The Hero as a Product of Female Mate Preferences

At the heart of the romance novel plot is a love story in which the heroine overcomes obstacles to identify, win the heart of, and ultimately have a long-term relationship with the one man who is right for her. The emotional focus of a romance is on love and commitment (often seen through the "taming" of the hero) with the goal the creation of a permanent bond with the heroine's ideal mate, one who is strong, yet caring and committed (Radway 1984). Unsurprisingly, with this focus on long-term mating, the romance hero is a reflection of female long-term mate preferences. Anthropologist April Gorry (1999) analyzed the descriptions of the heroes of 45 romance novels, nominated for their excellence by at least three romance readers or writers. In almost all these stories, the hero was older than the heroine by an average of seven years. Heroes were always described as taller than the heroine. The heroes' physical appearance was most frequently described as muscular, handsome, strong, large, tanned, masculine, and energetic. Gorry (1999) also reported romance heroes to exhibit cues of physical and social "competence." They were capable of handling whatever situations arose. Heroes were described as sexually bold, calm, confident, and impulsive. In most of the novels, the hero was described as "intelligent." All these traits reflect female preferences for males of good genetic quality who will be good protectors and providers. Romance writer Robyn Donald (1992), with an instinctive grip of the evolutionary underpinnings of the genre, sums up the hero's appeal:

> Until very recently in our historic past, strong, successful, powerful men had the greatest prospects of fathering children who survived. If a woman formed a close bond with a man who was sensible, competent and quick-witted, one high up in the family or tribal pecking order, a man with the ability to provide for her and any children she might have, the chances of her children surviving were greater than those of a woman whose mate was inefficient. (82)

The importance of a high-mate-quality hero is also revealed in research analyzing the titles of romance novels (Cox and Fisher 2009) as well as the development of hero and heroine (Fisher and Cox 2010). Cox and Fisher (2009) and Fisher and Cox (2010) predicted that the titles of the novels would reveal evidence of women's mate preferences—in particular, that females have preferences for males who provide resources helpful in raising offspring, males willing to commit to a long-term relationship, and males who have genetic quality. An examination of the 20 most frequently mentioned occupations in the titles reveals 17 professions for the hero falling into two categories: resource based (doctor, chief executive officer, royalty) and athletic or protector (cowboys, lawmen, soldiers). The majority of heroes are well equipped to provide stable resources, and have sufficient genetic quality and physical prowess to serve as a protector (not only of the female and any offspring, but also of their resources). The hero falling in love with the heroine guarantees his long-term commitment to her—commitment and reproduction being two other themes reflected in the titles.

Love, Hate, and Brotherhood in the Squared Circle

At first glance, professional wrestling might seem like an odd choice for an essay on consilience and storytelling. It's an unusual mix of athletics, drama, and comedy, played out by a cast fueled by testosterone in front of an audience of, often, rowdy fans. With its scripted outcomes, it's not entertainment in the traditional sports sense, but rather entertainment in the tradition of theater and film, whether one sits in the audience at a live event or experiences it on television in the comfort of home. It consists of a cast of characters, both good and evil, with a wide range of storylines that can be as simple as the conflict over who is the best man to the strange group and family dynamics seen in a number of World Wrestling Entertainment (WWE) arcs. Some pop culture theorists have declared it masculine melodrama (Jenkins 2005) and focus on either the morality play of good and evil (Barthes 2005) or a masculine backlash against the politically correct. Salmon and Clerc (2005) have also argued for attention to be paid to the significant percentage of female fans and whether they react differently to certain storylines and characters.

Pro wrestling is a subject ripe for a consilient approach. The ubiquitous themes of pro wrestling are those are the heart of a majority of human stories, including competition between males for status and resources, mate choice, justice, cheater detection, and male coalitions, as well as solidarity and rivalry among kin. Wrestling provides us with the same cast of characters as Shakespeare—the hero, the villain, the love interest, and the comedic foil—and, like Shakespeare's characters, they experience suffering and sacrifice as well as trust and betrayal. In this discussion, I focus on the ancestrally relevant themes of male status competition and cooperation, sibling cooperation, and rivalry. I also consider sex differences in audience response. (For a more comprehensive evolutionary look at pro wrestling, see Salmon [2012b].)

Male–Male Competition and Sexual Selection

The topic of male–male competition and sexual selection is one that has been well studied by evolutionary biologists and psychologists. In many species, males invest less in their offspring than females, and females tend to be choosy when it comes to mating because of their greater reproductive costs (Trivers 1985). This means that a male's reproductive success is largely the outcome of his ability to acquire mating opportunities that can occur by beating out other males for access to status or resources, or by exhibiting traits that females prefer, such as symmetrical features (Thornhill and Gangestad 2008). More successful males engage in risk-taking and aggression in the pursuit of status and resources (Buss and Schmitt 1993; Li 2007).

Pro wrestling is an aggressive performance in which physical competition between males takes place in front of an audience of males and females who are interested not only in who wins, but also in how and what he wins. Historically, pro wrestling was very focused on storylines involving two men fighting over the status of being the world champion. In the 1980s, when pro wrestling was experiencing a boom in popularity and revenue, this kind of status was epitomized by 16-time world champion Ric Flair, with his luxurious lifestyle of fancy ring attire, expensive suits and jewelry, Rolex watches, limousines, and a constant bevy of beautiful women by his side. Whether his role as champion was that of hero or villain, he was always portrayed as the man to beat.

But conflicts don't always revolve around these championships. Sometimes the status that is being fought over is that of being the toughest, a battle over respect, that strikes a chord with the young lower to middle-class male audience. One of the WWEs most popular stars in recent history built a reputation for the aggressive independent pursuit of victory. Stone Cold Steve Austin was one of a new generation of wrestlers who acted like the villain (rules don't matter), but nonetheless became the hero because he was seen as unwilling to give up or accept defeat. This reputation was cemented as part of a storyline involving a conflict over respect between him and Canadian wrestler Bret Hart, during which Austin lost a major match. At Wrestlemania 13 (a Pay-Per-View event), Austin had a no-disqualification submission match with Hart. The match ended when Austin, trapped in Hart's finishing hold, passed out from pain and blood loss. The victory was awarded to Bret, but Austin received thunderous applause from the fans for refusing to submit, and his never-surrender warrior attitude made him and his merchandise among the most popular and best-selling of all time for the WWE.

Social Alliances

When it comes down to it, whether a wrestler is a "nice guy" doesn't have a lot to do with being popular with the fans. When the hero (your favorite) breaks the rules, you justify it, and when someone we don't like does it, we call them the bad guy. But one of the easiest ways to turn a hero into a villain is to have a tag-team wrestler turn on his partner. From an evolutionary perspective, this should not be surprising. Males have always joined together for the purposes of hunting, warfare, and defense; humans are a cooperative, reciprocating species. Being socially successful involves a variety of skills or adaptations, including the ability to detect defectors, freeloaders, or cheaters (Cosmides 1989; Cosmides et al. 2010). Tag-team partners are portrayed as friends and allies, and trustworthiness is one of the most important characteristics in a coalition member. If your self-interests are not aligned, there's a nontrivial risk that your friend could become your foe. The successful maintenance of groups based on systems of alliances promoted the evolution of moral emotions. People want to see cheaters punished and admire the punishers. As a result, tag-team betrayals turn the betrayer from beloved favorite to hated villain almost instantly. For example, Hulk Hogan made a career out of being the cartoon hero ("Say your prayers; eat your vitamins") for over a decade, but that image became dated as the audience matured. The solution to growing fan indifference was to shock them by having the perennial good guy appear at an event where everyone assumed he was going to join forces with another fan favorite, Sting, who was under attack from a group of opponents. Instead, Hogan joined with the enemy (the nWo) in attacking Sting and got an incredibly strong reaction from the audience. The social and moral emotions that guide behavior, such as trust, fairness, and revenge, are at the center of many pro wrestling storylines—indeed, are at the heart of many pop culture works (Flesch 2008).

Brotherly Love and Hate

Now although legendary rivalries have been fought by unrelated males over titles and status, some of the most intense, those that capture the audience's heart and mind, have been those involving family, particularly the conflicts between brothers. Sibling relationships can often be quite complicated. Indeed, most sibling relationships can be characterized as having a fine line between love and hate, especially during childhood and adolescence (Cicirelli 1995). From an evolutionary perspective, this conflict stems from rivalry over limited resources (Salmon 2005), which often means competition over limited parental resources such as attention, time, and money in modern societies versus limited food and basic resources such as shelter in our evolutionary past. Investigations of proximate explanations have found the most common sources of sibling conflict to involve issues of relative power, self-interest (such as sharing of personal items), violation of rules (perceived immaturity and inappropriate behavior, for example), and interests outside the family (McGuire et al. 2000). But despite their conflicts, siblings can also be great assets. Older siblings can help care for younger ones (Cicirelli 1995; Bereczkei and Dunbar 2002). As adults, siblings can also be a source of social support (Connidis 1992; White 2001) and siblings are, next to parents, the only kin for which one is willing to make very costly sacrifices, such as donating a kidney (e.g., Neyer and Lang 2003).

Although there have been many sibling alliances and rivalries in wrestling, perhaps the most fantastical has been that between the Undertaker (portrayed by Mark Callaway) and his younger brother Kane (portrayed by Glenn Jacobs). The Undertaker's career began during a time in pro wrestling when the gimmick was all important. Wrestlers were portrayed much like comic book characters, and many stayed in character even outside the arena. Undertaker's skin was pale, his clothes were dark, he frequented graveyards, and his manager was named Paul Bearer. Some of his matches ended with his opponent placed in a casket left at ringside and there were even buried-alive matches with a temporary graveyard (imagine a mound of dirt next to a man-size gravelike hole) in the arena. After a split from his manager, Undertaker spent time as a fan favorite (as opposed to a scary villain), but Bearer soon returned with a surprise opponent for the Undertaker. It was his younger brother Kane, whom he had assumed was dead in a fire that had also killed their parents. Undertaker had a much more elaborate background story than most WWE characters, and that story was developed further with the appearance of his brother. Kane wore a mask and did not speak as a result of burns sustained in that fire. On televised events, Kane blamed his brother for the fire and his injuries, and a series of vicious matches ensued. Of course, there is a fine line between sibling love and hate, eventually circumstances changed and Kane and his brother buried their differences and united as the Brothers of Destruction.

Female Choice

Now the appeal of stories of male conflict to a young male audience is clear, but what draws the close to 25% of pro wrestling's audience (Salmon and Clerc 2005) that is female? Are women drawn to the same storylines? For the most part, it seems that well-crafted storylines appeal pretty equally to male and female audience members. Cross-gender appeal is evident in responses to the Undertaker/Kane saga as well as in responses to the usual conflicts over titles. However, men and women often respond to individual wrestlers in predictable ways. For example, given the relevance of male competition, one might predict that men would tend to favor the strongest, most dominant wrestlers, whereas women might favor wrestlers based on their mate choice appeal. WWE wrestler Brock Lesner has had a strong male following largely because of his obvious strength and power. He looks as if he could physically dominate pretty much anyone. However, he had a much smaller female following, probably because his ring personality is not very social or interesting and because he has relatively poor microphone skills. Dialogue skills are surprisingly important in pro wrestling; they are an excellent way to convey character and storyline detail, but they are also a source of information about personality traits such as kindness and humor that women use in assessing mate quality (Buss 1989; Thornhill and Gangestad 2008; Barclay 2010). One performer who has both a large male and a large female following is The Rock. A combination of athletic ability, incredible mic skills, a sarcastic sense of humor, social dominance, and attractiveness made him not only the "people's champion," but also won him the admiration of many.

The merchandise female fans purchase suggests how female mate choice preferences influence the popularity of wrestlers among female fans. The wrestlers favored by female fans tend to be facially attractive (showing signs of substantial testosterone exposure but not too much) as well as have attractive, strong bodies. They do not necessarily favor the largest wrestlers (like the Big Show) or the most muscular (like Lesner), but rather ones who look fit cardiovascularly (The Rock, Shawn Michaels, John Cena). Personality, as revealed through interviews and relationships with other wrestlers, also has an impact. Shows of friendship and solidarity—one wrestler coming to the rescue of another—are seen very favorably by female fans. A guy who wins all his matches, especially if he has a bland personality, may have a male following, but most female fans won't pay much attention. On the other hand, a guy who struggles at times yet wins in the end gives the women more information on which to judge him. Does he have drive and determination? The commitment to see things through? Does he show signs of attachment to others?

Based only on appearance and gimmick, the Undertaker appears an unlikely choice for a female favorite. He is quite aggressively masculine in appearance, closer to seven feet tall than six, and covered with tattoos. Conventionally facially attractive he is not. His powerful and intimidating presence has always been a huge draw to male fans, but he is tremendously popular with female fans. (See Salmon and Clerc [2005] for a discussion of his female fans and their romantic fantasies about him.) In a number of ways, he is the tall, dark, dangerous hero seen in Gothic romance. The ultimate protector with his size and strength, his somewhat predatory sexual appeal has been most obvious during his Lord of Darkness periods. Many of his promotional spots presented him as sensitive, tormented by tragedies in his past. Like many romance heroes, the initial impression is the dark and dangerous "cad," but later come glimpses of the kinder, gentler "dad" type.

Go Ahead, Make My Day and Band of Brothers: Fiction for Men

Traditional Approaches

Much of the academic focus on men and literature has examined what men read and, more broadly, why men read less than women. Some English scholars have explained the gender gap in reading as the product of socialization of how men should spend their leisure time (Tepper 2000) or the product of social status disparities (Kraaykamp and Dijkstra 1999). Others focus on how masculinity is represented in these genres consumed by men, including the conflict between the attraction of the noble outlaw and the man who defends his family and society from those who would prey on them (Chapman Peck 2008; Nichols 2008), the drive for individualism seen in the cowboy riding into town alone on his horse versus the local lawman and his family defending a town. Masculinity in these genres is also often discussed in terms of political ideology and the nation state in addition to gender roles such as being the "breadwinner" (Baker 2006; Mitchell 1996). Although these are all interesting interpretations of the stories, they don't particularly capture why these adventurous and exciting tales are so appealing to men young and old.

Ancestral Man in Modern Fiction

An adaptationist perspective would suggest that fiction that appeals particularly to men would be focused on the ancestral problems faced by men, including dealing with aggression and violence from other males, acquiring and defending social status, forming coalitions of allies, and playing the role of protector of family and the social group, including enforcing social rules (Daly and Wilson 1994; Price et al. 2002; Cosmides and Tooby 2005). Unsurprisingly, an examination of westerns and cop/military novels reveals a concentration on such problems. For example, Zane Grey's novels focus on life in the Old West, his heroes stoic and often lonely men who have to be self-reliant, but, like Buck Duane, the hero of *Lonestar Ranger*, want to do the right thing, and kill to rid towns of men who are corrupt and violent without cause.

Evidence suggests that warfare has been cross-culturally pervasive worldwide (Keeley 1996; Tooby and Cosmides 2010). Warfare is an activity pursued almost exclusively by men, with men the main victims, although women are also victimized. Men have adaptations designed to assess the competitive ability of themselves and others (Sell et al. 2009). They also have dispositions to form exclusively male cooperative coalitions and to experience intense bonds with their comrades while fighting against common enemies (Brown 1991; Browne 2007). This may account for the popularity and content of military stories, such as the true story bestseller *Bravo Two Zero* by former Special Air Service (SAS) officer Andy McNab (1993) This story focuses on a squad of eight British Special Forces men who are sent on a mission behind Iraqi lines and experience capture, torture, and the death of three of the squad members. Although billed as a story of men taken beyond the limits of human endurance, it is largely a tale of successful survival, as the authors writes at the end:

> As my stress-test score showed, I'm not emotionally affected by what happened. I certainly don't have nightmares. We are big boys and we know the rules we play by.... And as for the people who interrogated me, if

> I met any of them in the street tomorrow and thought I could get away with it, I'd slot them. (412)

It is a view of male conflict that doesn't hide how ugly and painful the violence of war can be, but it also explores how men might cope with these challenges.

Mack Bolan, also known as "The Executioner," is the fictional star of more than 600 novels with sales of more than 200 million copies (Thomas 1995). The character was created by Don Pendleton, who wrote 39 novels featuring Bolan, starting in 1969, before selling the rights to Gold Eagle Books, which continues to publish at least a dozen Mack Bolan novels every year. Interestingly, Gold Eagle is owned by Harlequin Books, the largest publisher of romance novels worldwide. A quick review of the titles illustrates the conflict/competition aspect of these stories: *Battle Plan, Battle Ground, War Hammer, Shifting Target, Lethal Agent, Death Force, Blood and Fire, Patriot Gambit,* and *Rolling Death*. Bolan started as a military sniper and, as the stories have accumulated, has focused on waging war against the mafia, KGB, and terrorist organizations. Brand is portrayed as a man of honor and a strong enforcer of justice. *Rolling Death* (1999) in which he brings down a Miami drug lord) has a preface with a quotation from the character: "Everyone who doesn't obey the law is fair game. Drug lords, gunmen, even the chemists—all who are laboring to further poison America" (p. 5). He is very much portrayed as the costly punisher, willing to risk his life to impose costs on those who threaten the social group. He is portrayed as a "man's man." When he isn't engaging in violence to bring down the bad guys, he's engaging in sex or extreme sports like white water rafting. Yet he's also portrayed as well read. The statement quoted from the preface is a response to a statement by Theodore Roosevelt concerning obedience to the law. Each of the novels begins with a quotation followed by Bolan's interpretation of it. He is a character men can find themselves admiring, successfully facing historically male problems of male aggression, warfare, and how to protect and defend. It's not surprising that this character has been around for more than 40 years and is still going strong.

Consilience and Pop Culture

Although there are numerous books and articles written on the subject of popular culture, most have failed to address the most basic questions about the underlying appeal to men and women of these genres. An adaptationist perspective allows us to examine how these genres embody the essential problems faced by men and women over an evolutionary time frame. Although the written (and performed) fictional genres discussed here were not present in our ancestral environment, they trigger our adaptations for mating, for male–male competition, and for social living all the same.

Others have made a well-established case for a now-flourishing consilient literary scholarship. It is one that doesn't just focus on qualitative methods, but quantitative ones as well (Gottschall 2008; Strout et al. 2010; see also Chapter 12, this volume). In addition, a 2012 special issue of the *Review of General Psychology* was focused on evolutionary approaches to a range of pop culture products from television shows (Fisher 2012), pornography (Salmon 2012a), horror movies (Clasen 2012), and comedy (Kuhle 2012) to country music (Kurzban 2012). A theoretical approach to the study of popular culture rooted in an evolutionarily informed understanding of human nature is one that explains why these genres are so popular and why they are popular with particular audiences. Although interesting work has already begun, much remains to be done to develop a truly consilient field of pop culture studies.

References

Baker, Brian. 2006. *Masculinity in fiction and film: Represented men in popular genres 1945–2000.* New York: Bloomsbury Publishing.

Barclay, Pat. 2010. Altruism as a courtship display: Some effects of third-party generosity on audience perceptions. *British Journal of Psychology* 101: 123–135.

Barthes, Roland. 2005. The world of wrestling. Reprinted in N. Sammond, ed., *Steel chair to the head: The pleasure and pain of professional wrestling*, 23–32. Durham, NC: Duke University Press.

Bereczkei, Tamas., and Robin. I. M. Dunbar. 2002. Helpers-at-the-nest among Hungarian gypsies. *Current Anthropology* 43: 804–809.

Brown, Donald. E. 1991. *Human universals.* New York: McGraw-Hill.

Browne, Kingsley. R. 2007. *Co-ed combat: The new evidence that women shouldn't fight the nation's wars.* New York: Sentinel HC.

Brownmiller, Susan. 1976. *Against our will: Men, women and rape.* New York: Bantam.

Buss, David. M. 1989. Sex differences in human mate preferences: Evolutionary hypotheses tested in 37 cultures. *Behavioral and Brain Sciences* 12: 1–49.

Buss, David. M., Max. Abbott, Alois. Angleitner, Armen. Asherian, Angela. Biaggio, et al. 1990. International preferences in selecting mates: A study of 37 cultures. *Journal of Cross-Cultural Psychology* 21: 5–47.

Buss, David. M., and David. P. Schmitt. 1993. Sexual strategies theory: An evolutionary perspective on human mating. *Psychological Review* 100: 204–232.

Buss, David. M., Todd. K. Shackelford, Lee. A. Kirkpatrick, and Randy. J. Larsen. 2001. A half century of American mate preferences. *Journal of Marriage and the Family* 63: 491–503.

Camp, Deborah 1997. The role of the romance novel. In Rita. Gallagher and Rita. C. Estrada, eds., *Writing romances: A handbook by the romance writers of America*, 46–53. Cincinnati, OH: Writer's Digest Books.

Chapman Peck, Wendy. 2008. The romance of competence: Rethinking masculinity in the Western. *Journal of Popular Film and Television* 30: 206–219.

Cicirelli, Victor. G. 1995. *Sibling relationships across the life span*. New York: Plenum Press.

Clasen, Mathias. 2012. Monsters evolve: A biocultural approach to horror stories. *Review of General Psychology* 16: 222–229.

Connidis, Ingrid. A. 1992. Life transitions and the adult sibling ties: A qualitative study. *Journal of Marriage and the Family* 54: 972–982.

Cosmides, Leda. 1989. The logic of social exchange: Has natural selection shaped how humans reason? Studies with the Wason selection task. *Cognition* 31: 187–276.

Cosmides, Leda., Clark. Barrett, and John. Tooby. 2010. Adaptive specializations, social exchange, and the evolution of human intelligence. *Proceedings of the National Academy of Sciences of the United States of America* 107: 9007–9014.

Cosmides, Leda., and John. Tooby. 2005. Neurocognitive adaptations designed for social exchange. In David. M. Buss, ed., *The handbook of evolutionary psychology*, 584–627. New York: Wiley.

Courtiol, Alexandre., Michel. Ramond, Bernard. Godelle, and Jean-Baptiste. Ferdy. 2010. Mate choice and human stature: Homogamy as a unified framework for understanding mate preferences. *Evolution* 64: 2189–2203.

Cox, Anthony., and Maryanne. Fisher. 2009. The Texas billionaire's pregnant bride: An evolutionary interpretation of romance fiction titles. *Journal of Social, Evolutionary, and Cultural Psychology* 3: 386–401.

Daly, Martin., and Margo. Wilson. 1994. Evolutionary psychology of male violence. In J. Archer, ed., *Male violence*, 253–288. London, UK: Routledge.

Donald, Robyn. 1992. Mean, moody, and magnificent: The hero in romance literature. In Jayne Ann. Krentz, ed., *Dangerous men and adventurous women: Writers on the appeal of the romance*, 81–84. Philadelphia: University of Pennsylvania Press.

Fisher, Maryanne. 2012. Why who shot J.R. matters: *Dallas* as the pinnacle of human evolutionary television. *Review of General Psychology* 16: 200–207.

Fisher, Maryanne., and Anthony. Cox. 2010. Man change thyself: Hero versus heroine development in Harlequin romance novels. *Journal of Social, Evolutionary, and Cultural Psychology* 4: 305–316.

Flesch, William. 2008. *Comeuppance: Costly signaling, altruistic punishment and other biological components of fiction*. Cambridge, MA: Harvard University Press.

Garcia, Justin. R., Chris. Reiber, Sean. G. Massey, and Ann. M. Merriwether. 2012. Sexual hookup culture: A review. *Review of General Psychology* 16: 161–176.

Gorry, April. 1999. *Leaving home for romance: Tourist women's adventures abroad.* Unpublished PhD diss., University of California at Santa Barbara.

Gottschall, Jon. 2008. *Literature, science, and a new humanities.* New York: Palgrave Macmillan.

———. 2012. *The storytelling animal: How stories make us human.* New York: Houghton Mifflin Harcourt.

Grodal, Torben. 2009. *Embodied visions: Evolution, emotion, culture and film.* Oxford, UK: Oxford University Press.

Hazen, Helen. 1983. *Endless rapture: Rape, romance and the female imagination.* New York: Charles Scribner's Sons.

Hughes, Susan. M., and Gordon. G. Gallup. 2003. Sex differences in morphological predictors of sexual behavior: Shoulder to hip and waist to hip ratios. *Evolution and Human Behavior* 24: 173–178.

Jenkins, Henry, III. 2005. Never trust a snake: WWF wrestling as masculine melodrama. In Nicholas. Sammond, ed., *Steel chair to the head: The pleasure and pain of professional wrestling,* 33–36. Durham, NC: Duke University Press.

Keeley, Lawrence. H. 1996. *War before civilization.* New York: Oxford University Press.

Kenrick, Douglas. T., Edward. K. Sadalla, Gary. Groth, & Melanie R. Trost. 1990. Evolution, traits, and the stages of human courtship: Qualifying the parental investment model. *Journal of Personality* 58: 97–116.

Kraaykamp, Gerbert., and Katinka. Dijkstra. 1999. Preferences in leisure tile book readings: A study on the social differentiation in book reading for the Netherlands. *Poetics* 26: 203–234.

Kuhle, Barry. X. 2012. It's funny because it's true (because it evokes our evolved psychology). *Review of General Psychology* 16: 177–186.

Kurzban, Robert. 2012. Cheatin' hearts and loaded guns: The high fitness stakes of country music lyrics. *Review of General Psychology* 16: 187–191.

La Cerra, Mary. M. 1994. *Evolved mate preferences in women: Psychological adaptations for assessing a man's willingness to invest in offspring.* Unpublished PhD diss., University of California at Santa Barbara.

Li, Norman. P. 2007. Mate preference necessities in long- and short-term mating: People prioritize in themselves what their mates prioritize in them. *Acta Psychologica Sinica* 39: 528–535.

McGuire, Shirley., Beth. Manke, Afsoon. Eftekhari, and Judy. Dunn. 2000. Children's perceptions of sibling conflict during middle childhood: Issues and sibling (dis) similarity. *Social Development* 9: 173–190.

McNab, Andy. 1993. *Bravo Two Zero.* London: Bantam Press.

Mitchell, Lee. C. 1996. *Westerns: Making the man in fiction and film.* Chicago: University of Chicago Press.

Modleski, Tania. 1980. The disappearing act: A study of Harlequin romances. *Signs* 5: 434–448.

Mussell, Kay. 1984. *Fantasy and reconciliation: Contemporary formulas of women's romance fiction*. Westport, CT: Greenwood Press.

Neyer, Franz. J., and Frieder. R. Lang. 2003. Blood is thicker than water: Kinship orientation across adulthood. *Journal of Personality and Social Psychology* 84: 310–321.

Nichols, Mary. P. 2008. Revisiting heroism and community in contemporary westerns: *No Country for Old Men* and *3:10 to Yuma*. *Perspectives on Political Science* 37: 207–216.

Pendleton, Don. 1999. *Rolling death*. New York: Worldwide.

Price, Michael. E., Leda. Cosmides, and John. Tooby. 2002. Punitive sentiment as an anti-free rider psychological device. *Evolution and Human Behavior* 23: 203–231.

Radway, Janice. 1984. *Reading the romance: Women, patriarchy, and popular literature*. Chapel Hill: University of North Carolina Press.

Rhodes, Gillian. 2006. The evolutionary psychology of facial beauty. *Annual Review of Psychology* 57: 199–226.

Roney, James. R., Katherine. N. Hanson, Kristina. M. Durante, and Dario. Maestripieri. 2006. Reading men's faces: Women's mate attractiveness judgments track men's testosterone and interest in infants. *Proceedings of the Royal Society of London Series B* 273: 2169–2175.

Salmon, Catherine. 2012a. The pop culture of sex: An evolutionary window on the worlds of pornography and romance. *Review of General Psychology* 16: 152–160.

———. 2012b. The world of sports entertainment: Evolutionary drama in the squared circle. *Evolutionary Review* 3: 108–115.

Salmon, Catherine. A. 2005. Parental investment and parent–offspring conflict. In D. M. Buss, ed., *Handbook of evolutionary psychology*, 506–527. Hoboken, NJ: Wiley.

Salmon, Catherine., and Susan. Clerc. 2005. Ladies love wrestling too: Female wrestling fans online. In Nicholas. Sammond, ed., *Steel chair to the head: The pleasure and pain of professional wrestling*, 167–191. Durham, NC: Duke University Press.

Salmon, Catherine., and Donald. Symons. 2001. *Warrior lovers: Erotic fiction, evolution, and female sexuality*. London, UK: Weidenfeld and Nicolson.

Sell, Aaron., Leda. Cosmides, John. Tooby, Daniel. Sznycer, Christopher. von Rueden, and Michael. Gurven. 2009. Human adaptations for the visual assessment of strength and fighting ability from the body and face. *Proceedings of the Royal Society of London Series B*, 276: 575–584.

Strout, Sarah., Maryanne. Fisher, Daniel. Kruger, and Lesley-Ann. Steeleworthy. 2010. Pride and prejudice or children and cheating? Jane Austen's representations of female mating strategies. *Journal of Social, Evolutionary, and Cultural Psychology* 4: 317–331.

Tepper, Steven. 2000. Fiction reading in America: Explaining the gender gap. *Poetics* 27: 255–275.

Thomas , Robert Jr. 1995. Don Pendleton, 67, writer who spawned a genre. *New York Times*, October 28. http://www.nytimes.com/1995/10/28/us/don-pendleton-67-writer-who-spawned-a-genre.html

Thornhill, Randy., and Steven. W. Gangestad. 2008. *The evolutionary biology of human female sexuality*. New York: Oxford University Press.

Tooby, John., and Leda. Cosmides. 2010. Groups in mind: The coalitional roots of war and morality. In Henrik. Hogh-Oleson, ed., *Human morality and sociality: Evolutionary and comparative perspectives*, 191–234. New York: Palgrave MacMillan.

Tooby, John., and Leda. Cosmides. 2001. Does beauty build adapted minds? Toward an evolutionary theory of aesthetics. *SubStance* 30: 6–27.

Trivers, Robert. L. 1972. Parental investment and sexual selection. In B. Campbell, ed. *Sexual selection and the descent of man: 1871–1971*, 136–179. Chicago: Aldine.

Trivers, Robert. 1985. *Social evolution*. Menlo Park, CA: Benjamin/Cummings.

White, Lynn. 2001. Sibling relationships over the life course: A panel analysis. *Journal of Marriage and the Family* 63: 555–568.

11 TERRIFYING MONSTERS, MALEVOLENT GHOSTS, AND EVOLVED DANGER MANAGEMENT ARCHITECTURE

A CONSILIENT APPROACH TO HORROR FICTION

Mathias Clasen

When darkness falls, the monsters come out. That's a truism of prehistory as well as horror fiction. Life in prehistoric times was dangerous. Our ancestors have, for millions of years, been under threat from mammalian predators, venomous animals, hostile conspecifics, and invisible pathogens (Barrett 2005; Hart and Sussman 2008). The selection pressure from these dangers has resulted in a species-typical cognitive architecture or hardware for danger management (Öhman and Mineka 2001; Eilam et al. 2011; Woody and Szechtman 2011), and horror fiction runs on this hardware (Clasen 2010b, 2012c). Thus, an understanding of human evolutionary history contributes crucially to explaining the structure and function of horror stories. I argue for an approach to horror fiction that builds on the evolutionary social sciences. In recent decades, scientists and scholars have made considerable progress in delineating human nature (Pinker 2002) and in locating the study of culture within a naturalistic framework integrated with the evolutionary social sciences and, ultimately, the life sciences (Carroll 2004; Boyd 2009; Carroll 2011; Slingerland and Collard 2011). By assimilating that research, a consilient approach to horror fiction offers a corrective to theoretically flawed approaches that have dominated horror studies in recent decades, especially psychoanalysis and the various forms of "blank slate" political ideology (Clasen 2010b, 2012c).

Horror fiction takes its name from an emotion. It is affectively defined—according to intended audience response—just like comedy. Those formal elements that typically invoke negative emotion (fear, terror, anxiety, disgust) in their audiences are found in all narrative and visual media, and figure in disparate genres. But horror fiction proper has as a primary and irreducible goal the evocation of negative emotion in its audience. The theatrical trailer for the 2007 independent hit movie *Paranormal Activity* alternated shots of a theater audience watching the film with scenes from the film itself. In grainy, green-tinged night-vision images, the trailer showed an audience recoiling in fear, covering their eyes, laughing nervously, and screaming in terror. When *The Exorcist* (1973) opened in Danish theaters in 1974, cinema operators handed out stickers to ticket buyers with the words: "I have seen THE EXORCIST, the movie the whole world is talking about. DO YOU DARE WATCH IT?" (Clasen 2012d, my translation). The consumers of horror expect to be scared.

One needs to understand negative emotions such as fear and anxiety to understand imaginative constructions that are designed to evoke such emotions. The universes of supernatural horror literature, films, and video games feature fearsome monsters or otherworldly forces of evil preying on human victims—preying on the body or on the soul, or both. Horror stories stage a virtual reenactment of a timeless struggle for existence in the face of grave natural dangers and baffling forces in the environment. Many modern horror stories embellish and exaggerate the historically prevalent natural dangers posed by predators, toxic organisms, invisible pathogens, and hostile conspecifics. Embellishment and exaggeration enable the stories to target cognitive danger management machinery all the more effectively. Physically harmless yet spiritually odious monsters such as malevolent ghosts are less obviously reflections of ancestral dangers than are predatory monsters such as werewolves, but malevolent ghosts also spring from the workings of adapted minds trying to make sense of the world by positing meaningful causal relations and disembodied intentional agents where none exists.

Horror fiction lets us experience worst-case scenarios in decoupled mode; it allows us to face the worst anyone can imagine, but without serious risk or even cost. An emerging line of thinking in cognitive and evolutionary literary study regards fiction as a cultural technology with a primary function that is imaginative simulation—experiments in the lab of the mind. Raymond Mar and Keith Oatley (2008) argue that the main function of fiction is social simulation (see also Gottschall 2012). I argue that the main function of horror fiction is *emotional* simulation (see also Grodal 2009; Eitzen 2010; Clasen 2012b).

Horror Fiction and the Elicitation of Negative Emotion

Emotions are adaptations in the biological sense. They are evolved mechanisms that orchestrate complex responses adaptively appropriate to their domains

(Cosmides and Tooby 2000), so that, for example, fear directs our attention toward the fear elicitor while readying the body for action (confrontation or evasion). Among other functions, fear redirects energetic resources from the digestive system—irrelevant if one is being attacked by a hungry predator—to the muscle groups involved in combat and urgent flight. Öhman and Mineka (2001) posit an evolved "fear module" that is "a device for activating defensive behavior . . . and associated psychophysiological responses and emotional feelings to threatening stimuli" (485). Because of its deep history in vertebrate evolution, the human fear module is particularly sensitive to ancestrally relevant fear objects such as predators and venomous animals like some snakes and spiders (Marks and Nesse 1994; Öhman and Mineka 2001).

According to Robert Plutchik (2001), "an emotion is not simply a feeling state. Emotion is a complex chain of loosely connected events that begins with a stimulus and includes feelings, psychological changes, impulses to action and specific, goal-directed behavior" (345–346). The audience reaction evoked by horror fiction follows Plutchik's chain of events but suppresses the "specific, goal-directed behavior" normally elicited by negative stimuli. A reader or a viewer perceives a stimulus, such as a representation of a horrible monster, that evokes certain domain-specific feelings and causes psychological and physiological change, and ignites impulses to action. The audience reacts to representations of horrible creatures and threatening situations with negative emotion, yet the primal emotions evoked by horror fiction are likely modulated by "higher," evolutionarily younger cognitive faculties with neural underpinnings in the prefrontal cortex (Straube et al. 2010), so that most behavioral effects of such emotions are repressed. We know it's fiction; we know there's no real danger. People rarely run screaming from movie theaters or attack a Stephen King novel in self-defense. More subtle behavioral responses do occur, such as squirming in one's seat, covering one's eyes, scanning the page anxiously to see what happens next, looking over one's shoulder, and so on, but the behavioral component of the fight-or-flight response is normally absent in the experience of a fictional horror story. In other words, horror fiction is experienced in decoupled mode (Tooby and Cosmides 2001).

Horror fiction runs on cognitive machinery for danger management. All horror stories attempt to evoke negative emotion. Some, particularly in film and video gaming, rely on shock effects to elicit startle responses. But narrative horror also relies more subtly on establishing an atmosphere of dread and impending danger. In *The Terror*, Dan Simmons's (2007) novel of a doomed 19th-century arctic expedition, a giant white bear with supernatural capacities plays a central role and looms over the whole story. Yet, like Count Dracula in Stoker's classic novel (Clasen 2012a), the monster is given very little stage time. Readers and characters never know when the monster will strike. In horror films, the monster is typically shown onscreen only after a lengthy buildup in which formal elements work together to create an atmosphere of impending danger. In *The Exorcist* (1973), the massive disturbance caused by the possession of Regan McNeil by the demon Pazuzu is prefigured with weird scratching noises in the attic and a candle-lit nighttime expedition to the attic by the mother. Dreamlike sequences of almost subliminal demonic horror imagery further manipulate audience response and expectations.

The anxiety evoked by horror stories is closely related to fear, but whereas fear is the adaptive response to imminent danger (such as from a predator or a hostile conspecific), anxiety is the adaptive response to distant, potential, or abstract threats (Eilam et al. 2011). The characteristic behavioral output of anxiety involves probing the environment and gathering information, rather than avoidance (Eilam et al. 2011). Thus, establishing an atmosphere of dread is an effective strategy for commanding the audience's attention and keeping them focused on the action. This strategy is centrally dependent on evolved properties of the human central nervous system.

Psychological Effects and Adaptive Function of Horror Fiction

In the horror film *Scream* (1996), a character is on the phone with a crazed killer who appears to hide nearby. The character nervously calls out, "Who's there?" The crazed killer, Ghostface, flippantly answers: "Never say 'who's there?' Don't you watch scary movies? It's a death wish. You might as well come out to investigate a strange noise or something." If one were looking to horror fiction for bits of practical advice, this might be about the best one could anticipate receiving. Horror fiction can nonetheless still serve an adaptive function: it gives us experience with negative emotion at levels of intensity not safely come by in real life. It thus allows us to incorporate the imagination of danger into our total imaginative universe. One might never find oneself face-to-face with a crazed chainsaw killer or a giant white bear—or even a mundane low-life assailant or an ordinary brown bear; but, as Paul Bloom (2010) points out, even unrealistic zombie stories can serve as "useful practice for bad times, exercising our psyches for when life goes to hell" (193–194). Danger is an inescapable part of all life on earth. No one gets out alive, and in ancestral environments probably very few died peacefully of sheer old age. Even now, we live safely in highly controlled urban environments only on condition of being at least subliminally aware that the world contains many potential dangers against which we must be on guard. In dangerous environments, staying alert means staying alive (Wilson 1984).

Dramatized stories about horrible danger can inflate our sense of insecurity; there is empirical evidence to suggest that scary stories sensitize us to real-life danger, sometimes debilitatingly so. Even stories about outlandish, supernatural monsters can instill or exacerbate a sense of insecurity and urge us toward more cautious behavior. One may feel increased anxiety and heightened vigilance over long periods after exposure to a particularly effective horror story. Joanne Cantor (2004) has researched the "lingering effects of frightening media" (283). In a recent study, she collected and quantified 530 reports from individuals whom she had asked to write about fright reactions to media presentations. Strikingly, 91% of the respondents described negative reactions to fictional media, rather than to news or documentary presentations. One respondent describes how he watched the horror film *Poltergeist* (1982) at a young age and refused to sleep with an open closet door in the room for several months (Cantor 2004, 289). Many other respondents claimed to have suffered from nightmares and other sleep disturbances for months after seeing the film (Cantor 2004, 290). Several respondents described the effects of *The Blair Witch Project* (1999), a low-budget, independent production detailing a group of young people lost and preyed on in dark and strange woods (Cantor 2004, 293). Some refused to go camping or be in the woods. For several days after seeing the film, another felt compelled to leave all the lights on in her apartment. In another study, Cantor and her colleague Becky Omdahl (1991) exposed children (kindergarten through sixth grade) to "scary media presentations" to see whether fictional presentations of realistic life-threatening events would influence the children's risk assessment. They found that children who had been exposed to a "dramatized depiction of a deadly house fire from *Little House on the Prairie* increased their self-reports of worry about similar events in their own lives." Moreover, the children "were also less interested in learning to build a fire in a fireplace" than children who had not seen the scene or who had seen a nondramatic scene involving fire (Cantor 2009, 289).

Horror stories work by tapping into highly conserved danger management adaptations. They feature fear-inducing monsters and anxiety-inducing situations. Strikingly, such stories often depict dangers that are not real dangers in modern environments, but the psychological responses to these often unrealistic dangers are real psychological responses, and because they enable us to engage with the imagination of danger and to incorporate it into our imaginative and thus emotional system, the consumption of horror stories may be adaptive.

Immersing oneself into the bleak and threatening imaginative universes of horror fiction, in short, may prepare us for when "life goes to hell."

Imaginary Predators and Horrors from Beyond the Grave

When the American Centers for Disease Control and Prevention (CDC) launched their "Preparedness 101: Zombie Apocalypse" campaign in 2011, they seemed to realize the fascination the zombie commands and the pleasure people take in imagining worst-case scenarios (Clasen 2010a). The CDC used the zombie figure as a vehicle to strengthen emergency vigilance because zombies and stories about the zombie apocalypse are simply more interesting than dry facts about hazard preparedness. The notorious zombie campaign began as a joke, but as the CDC notes on its website, it soon proved effective in engaging people who wouldn't otherwise pay attention to the CDC's preparedness messages: "We continue to reach and engage a wide variety of audiences on all hazards preparedness via Zombie Preparedness; and as [CDC] director, Dr. Ali Khan, notes, 'If you are generally well equipped to deal with a zombie apocalypse, you will be prepared for a hurricane, pandemic, earthquake, or terrorist attack'" (Centers for Disease Control and Prevention 2011).

People are fascinated with dangerous monsters such as zombies because they are supernormal predators—anomalous embodiments of the instruments of death (Clasen 2010b). They are hostile and powerful. They have the means to inflict physical or spiritual damage. The giant white bear in Simmons's (2007) *The Terror* has zoologically implausible supernatural qualities, but it is clearly conceptualized as a predator. The monster is "a dark mass of hair and muscle and sunset-tinted claws and a faint gleam of teeth beyond anything . . . in mankind's racial memory of its many predators" (681). It has the anatomical characteristics of mammalian predators—claws, teeth, and muscle—and is exaggerated along dimensions of fearfulness.

We are naturally alert to dangers in the environment and curious to learn about them. Such curiosity is adaptive (Kruuk 2002; Barrett 2005) because it allows us to calibrate our "predatory defense system" (Öhman 1985), a system that needs environmental input, or learning, to function properly. "Learning," according to Öhman and Susan Mineka (2001), "is an evolutionarily derived adaptation to cope with environmental changes that occur within the life span of individuals and allows individual organisms to tailor their behavior to the specific environmental niche they occupy" (487). Öhman (1985) writes that "evolutionary economy has left to environmental influences to inscribe the exact characteristics of dangerous predators" (128–129) and invokes Martin

Seligman's (1971) notion of "prepared learning" to explain why certain stimuli are vastly easier to learn to fear than others. People acquire fear of ancestrally dangerous entities such as snakes and spiders much more easily than they do of things that are dangerous in present-day environments but that exerted no selection pressure in ancestral ones, such as motorized lawnmowers (LoBue 2013). Learning to fear objects and organisms, in other words, occurs within a possibility space constrained by evolutionary forces.

Some horror monsters are only minimally anomalous and clearly reflect ancestral dangers, such as the super-size spiders in *The Mist* (2007), whereas others defy basic natural laws in their strangeness and pose no direct physical threat. The malevolent ghost, an archetypal figure of horror fiction, is a case in point. The ghost typically does not threaten directly with physical harm yet is feared by the protagonists of the story. As Joseph Carroll points out:

> [Ghosts] are felt to intermingle death into life, violating categories in ways that threaten life and well-being among the living. A ghostly touch is like the touch of death. It is almost always cold, like a little foretaste of mortal disease, and indeed, like an infection from death. The realm of death threatens constantly to overwhelm life. And eventually, it always does. (pers. comm., July 25, 2012)

Carroll's observations run parallel with suggestions in Joe Hill's horror novel *Heart-Shaped Box* (2007). Hill's protagonists are haunted by a malevolent ghost who threatens to invade their minds and force them to commit suicide. The ghost shows itself to the protagonists and also communicates via radio broadcasts and even e-mail messages. In one such message, the ghost writes: "you will die anyone who gets too close will be infected with the death on you . . . we are infected together we will be in the death-hole together" (Hill 2007, 78). Yet more than fear of death, ghosts evoke a fear of premature and profoundly unfair death. There's no way to have a fair fight with a ghost. One can combat an angry werewolf or a hungry zombie, but, as Hill (2007) writes, the "ghosts always caught up eventually, and there was no way to lock the door on them" (202).

The idea of a ghost is predicated on mind–body dualism, which is not just a metaphysical standpoint but a fundamental design characteristic in human cognitive architecture, probably an artifact of an evolved Theory of Mind capacity—the intuitive understanding that other people have beliefs that may differ from one's own (Bloom 2004). People assume nonreflectively that other people have an inner intangible mental life, a nonphysical existence or spirit, no matter what neurobiologists tell us about the neural underpinnings of consciousness. One does not have to jump very far to reach the conclusion that the spirit just may stick around after the body dies (Boyer 2001). Moreover, an innate tendency to overattribute agency to apparently inexplicable events strengthens belief in ghosts (Atran and Norenzayan 2004; Clasen 2012c). Such cognitive biases keep folk-belief in ghosts alive, and popular culture has eagerly embraced the concept of the ghost as a highly salient figure, an attention-demanding messenger from beyond. Theory of Mind evolved as a result of selection pressures stemming from social organization, the need to infer other people's intentions from their behavior. Overattribution of agency evolved as a result of selection pressures stemming from predation—the simple rule that it's better to be safe than sorry. If one hears a strange rustle in the woods or an unexpected noise in the dead of night, it's better to react defensively—to assume something is out there—than to dismiss the noise if it turns out to be a cue of danger.

There is a direct causal chain from ancestral selection pressures to the prevalence of malevolent ghosts and fearsome monsters in horror fiction. Adapted minds project revenants and other supernatural monstrosities into the world. Evolved cognitive biases disposing us to believe in supernatural agents embellish and exaggerate actual predators. Because humans are disposed to be attuned to dangerous organisms, even imaginary monsters catch our attention. A theory that attempts to explain the prevalence and the psychological function of imaginary monsters and horror stories needs to take into account evolved cognitive architecture, and that is the promise of a consilient theory of horror fiction.

A Consilient Framework

These are exciting times in intellectual history. We are finally developing the theoretical tools, the empirical foundation, and the conceptual resolution necessary for unearthing the causal links that run from evolutionary biology to the most refined products of the human imagination, and back again. Consilient literary scholarship has shown its explanatory power and critical utility, and some few literary scholars are now embracing quantitative analytic tools from the social sciences and applying them to specifically literary questions (Chapter 12, this volume). As Carroll and his colleagues show, there is no good epistemological reason for keeping the study of literature quarantined from quantitative methodology (see also Gottschall 2008; van Peer et al. 2012). The adaptive function of horror fiction, for example, is an empirical question, and there is no need to wait for the zombie apocalypse to figure out the answer.

I have, in previous work, shown how evolutionary thinking can inform interpretive criticism of particular horror texts (Clasen 2010c, 2011, 2012a) and have in this chapter delineated a consilient theory of horror fiction and supernatural monsters, a theory that is lodged in the evolutionary social sciences. Evolved dispositions to be captivated by dangerous agents explain the prevalence of monsters in fiction, and the design specs of cognitive danger management machinery explain how and why horror fiction works. But much work remains to be done in the field of consilient horror study. This is a moment of opportunity for horror scholars to embrace tools and findings from neighboring sciences. Situating the study of horror fiction within an internally consistent, naturalistic framework will allow us finally to break open the paradox of horror—the strange fact that many people are attracted to fictional stories designed manifestly to terrify them.

References

Atran, Scott, and Ara Norenzayan. 2004. Religion's evolutionary landscape: Counterintuition, commitment, compassion, communion. *Behavioral and Brain Sciences* 27: 719–730.

Barrett, H. Clark. 2005. Adaptations to predators and prey. In David M. Buss, ed., *Handbook of evolutionary psychology*, 200–223. New York: Wiley.

Bloom, Paul. 2004. *Descartes' baby: How the science of child development explains what makes us human*. New York: Basic Books.

———. 2010. *How pleasure works: Why we like what we like*. London: Vintage Books.

Boyd, Brian. 2009. *On the origin of stories: Evolution, cognition, and fiction*. Cambridge, MA: Harvard University Press.

Boyer, Pascal. 2001. *Religion explained: The evolutionary origins of religious thought*. New York: Basic Books.

Cantor, Joanne. 2004. "I'll never have a clown in my house": Why movie horror lives on. *Poetics Today* 25: 283–304.

———. 2009. Fright reactions to mass media. In Jennings Bryant and Mary Beth Oliver, eds., *Media effects: Advances in theory and research*, 3rd ed., 287–303. New York: Routledge.

Cantor, Joanne, and Becky L. Omdahl. 1991. Effects of fictional media depictions of realistic threats on children's emotional responses, expectations, worries, and liking for related activities. *Communication Monographs* 58: 384–401.

Carroll, Joseph. 2004. *Literary Darwinism: Evolution, human nature, and literature*. New York: Routledge.

———. 2011. *Reading human nature: Literary Darwinism in theory and practice*. Albany, NY: SUNY University Press.

Centers for Disease Control and Prevention. *Zombie Preparedness, 2011*. CDC, Office of Public Health and Preparedness Response. http://www.cdc.gov/phpr/zombies.htm [Online].

Clasen, Mathias. 2010a. The anatomy of the zombie: A bio-psychological look at the undead other. *Otherness: Essays and Studies* 1: 1–23.

———. 2010b. The horror! The horror! *Evolutionary Review: Art, Science, Culture* 1: 112–119.

———. 2010c. Vampire apocalypse: A biocultural critique of Richard Matheson's *I Am Legend. Philosophy and Literature* 34: 313–328.

———. 2011. Primal fear: A Darwinian perspective on Dan Simmons' *Song of Kali. Horror Studies* 2: 84–104.

———. 2012a. Attention, predation, counterintuition: Why Dracula won't die. *Style* 43: 396–416.

———. 2012b. "Can't sleep, clowns will eat me": Telling scary stories. In Dirk Vanderbeke and Carsten Gansel, eds., *Telling stories: Literature and evolution*, 324–346. Berlin, Germany: de Gruyter.

———. 2012c. Monsters evolve: A biocultural approach to horror stories. *Review of General Psychology* 19: 222–229.

———. 2012d. *Monstre.* Aarhus, Denmark: Aarhus University Press.

Cosmides, Leda, and John Tooby. 2000. Evolutionary psychology and the emotions. In Michael Lewis and Jeanette M. Haviland-Jones, eds., *Handbook of emotions*, 2nd ed., 91–115. New York: Guilford.

Eilam, David, Rony Izhar, and Joel Mort. 2011. Threat detection: Behavioral practices in animals and humans. *Neuroscience and Biobehavioral Reviews* 35: 999–1006.

Eitzen, Dirk. 2010. *The fun of fear: horror, suspense, and Halloween.* The Society for Cognitive Studies of the Moving Image. http://scsmi-online.org/forum/the-fun-of-fear-horror-suspense-and-halloween [Online].

Gottschall, Jonathan. 2008. *Literature, science, and a new humanities.* New York: Palgrave Macmillan.

———. 2012. *The storytelling animal: How stories make us human.* New York: Houghton Mifflin Harcourt.

Grodal, Torben. 2009. *Embodied visions: Evolution, emotion, culture, and film.* New York: Oxford University Press.

Hart, Donna, and Robert W. Sussman. 2008. *Man the hunted: Primates, predators, and human evolution*, expanded ed. Boulder, CO: Westview.

Hill, Joe. 2007. *Heart-shaped box.* London: Orion.

Kruuk, Hans. 2002. *Hunter and hunted: Relationships between carnivores and people.* Cambridge: Cambridge University Press.

LoBue, Vanessa. 2013. What are we so afraid of? How early attention shapes our most common fears. *Child Development Perspectives* 7: 38–42.

Mar, Raymond A., and Keith Oatley. 2008. The function of fiction is the abstraction and simulation of social experience. *Perspectives on Psychological Science* 3: 173–192.9

Marks, Isaac M., and Randolph M. Nesse. 1994. Fear and fitness: An evolutionary analysis of anxiety disorders. *Ethology and Sociobiology* 15: 247–261.

Öhman, Arne. 1985. Face the beast and fear the face: Animal and social fears as prototypes for evolutionary analyses of emotion. *Psychophysiology* 23: 123–145.

Öhman, Arne, and Susan Mineka. 2001. Fears, phobias, and preparedness: Toward an evolved module of fear and fear learning. *Psychological Review* 108: 483–522.

Pinker, Steven. 2002. *The blank slate: The modern denial of human nature.* London: Penguin.

Plutchik, Robert. 2001. The nature of emotions. *American Scientist* 89: 344–350.

Seligman, Martin. 1971. Phobias and preparedness. *Behavior Therapy* 2: 307–321.

Simmons, Dan. 2007. *The Terror.* London: Bantam Books.

Slingerland, Edward, and Mark Collard, eds. 2011. *Creating consilience: Integrating the sciences and the humanities.* New York: Oxford University Press.

Straube, Thomas, Sandra Preissler, Judith Lipka, Johannes Hewig, Hans- Joachim Mentzel, and Wolfgang H. R. Miltner. 2010. Neural representation of anxiety and personality during exposure to anxiety-provoking and neutral scenes from scary movies. *Human Brain Mapping* 31: 36–47.

Tooby, John, and Leda Cosmides. 2001. Does beauty build adapted minds? Toward an evolutionary theory of aesthetics, fiction and the arts. *SubStance* 30: 6–27.

van Peer, Willie, Frank Hakemulder, and Sonia Zyngier. 2012. *Scientific methods for the humanities.* Amsterdam: John Benjamins Publishing.

Wilson, Edward O. 1984. *Biophilia.* Cambridge, MA: Harvard University Press.

Woody, Erik, and Henry Szechtman. 2011. Adaptation to potential threat: The evolution, neurobiology, and psychopathology of the security motivation system. *Neuroscience and Biobehavioral Reviews* 35: 1019–1033.

Filmography

The Blair Witch Project, directed by Daniel Myrick and Eduardo Sánchez. 1999. Haxan Films.

The Exorcist, directed by William Friedkin. 1973. Warner Bros. Pictures.

The Mist, directed by Frank Darabont. 2007. Dimension Films/Darkwoods Productions/The Weinstein Company.

Paranormal Activity, directed by Oren Peli. 2007. Blumhouse Productions.

Poltergeist, directed by Tobe Hooper. 1982. Metro-Goldwyn-Mayer/SLM Production Group.

Scream, directed by Wes Craven. 1996. Dimension Films/Woods Entertainment.

12 AGONISTIC STRUCTURE IN CANONICAL BRITISH NOVELS OF THE NINETEENTH CENTURY

Joseph Carroll, Jonathan Gottschall, John A. Johnson, and Daniel Kruger

Historical and Methodological Context

Evolutionary Literary Study

Literature did not become the subject of an academic discipline until the last two decades of the 19th century and, until the 1940s, literary scholarship consisted chiefly of philological and historical scholarship and moralized aesthetic commentary (Abrams 1997; Graff 2007). During the 1930s, "The New Criticism" introduced methods for the intensive formal analysis of theme, tone, and style. During the late 1970s, "poststructuralism" or "postmodernism," spearheaded by the "deconstructive" philosophy of Jacques Derrida, produced a revolution in literary studies. Deconstruction identifies language or "discourse" as the primary constitutive material of human experience. In its political aspect, poststructuralism seeks to undermine traditionally dominant terms in social, psychological, and sexual binaries: ruling classes versus the oppressed, whites versus people of color, colonialists versus colonized peoples, mentally healthy people versus the insane, law-abiding citizens versus outlaws, males versus females, and heterosexuals versus homosexuals. In modern Western civilization, science is itself a dominant cultural value and is contrasted with terms such as *superstition, faith, ignorance, mysticism,* and *ideology.* In its epistemological aspect, poststructuralist theories of science seek to undermine the ideas of "truth" and "reality" through which science claims normative epistemic authority (Gross and Levitt 1994; Gross et al. 1996; Fromm 1997; Koertge 1998;

Sokal and Bricmont 1998; Weinberg 2001; Parsons 2003; Boghossian 2006; Smith 2006).

Before the poststructuralist revolution, humanists for the most part felt that their own kind of intellectual activity—scholarly, impressionistic, intuitive, and discursive—was fundamentally distinct from the activity of the sciences, both the physical and the social sciences. The "two cultures," as C. P. Snow (1993) designated them, were supposed to have different subject matters, to operate according to different rules, and to produce different kinds of knowledge. New Critics regarded literary texts as autonomous systems of meaning, independent of all external conditioning, either social or biographical. Poststructuralist theory expanded the notion of textual autonomy to include not just the isolated literary text, but also the whole textual universe—the world constituted by "discourse." The idea of cultural autonomy brings "standard social science"—that is, nonevolutionary social science—into partial alignment with poststructuralism, and during the 1990s, poststructuralist theory began to seep over into anthropology. Much standard social science nonetheless remains epistemologically distinct from poststructuralism. Even when social scientists reject the idea that genetically transmitted dispositions influence culture, most still regard scientific methodology as a medium of objective knowledge about a real world that exists independently of cultural and linguistic constructs.

During the past two decades, a growing body of literary scholars has assimilated research in the evolutionary human sciences. Variously known as "literary Darwinists" "biocultural critics" or "evolutionary literary scholars," these scholars have rejected the antirationalism of the poststructuralists and the blank-slate model of human nature that informs standard social science. They have rejected also the idea that science and the humanities form two distinct cultures, with different subject areas, different forms of knowledge, and different criteria of validity. In adopting the framework of evolutionary social science, the literary Darwinists adopt an overarching rationale for the integration of all disciplines under the canons of scientific criteria of epistemic validity. They believe that nature forms a unified causal network and that science provides an integrated understanding of that network. Nature forms a nested hierarchy in which more elementary causal forces constrain the organization of phenomena at higher levels. Thus, causal forces in physics constrain chemical phenomena, the causal forces in chemistry constrain biological phenomena, the causal forces in biology constrain human psychology, and the causal forces in psychology constrain all cultural products, including literature and the other arts.

In *Consilience: The Unity of Knowledge,* Edward O. Wilson (1998) identifies the humanities as the last frontier for bringing all possible phenomena within the scope of scientific understanding. Unlike poststructuralist theorists of science, literary scholars who concur with Wilson do not seek to assimilate science to the theory of "discourse." Instead, they seek to bring all discursive and imaginative activity within the scope of subjects accessible to science. Like the majority of literary authors and theorists from the time of Aristotle until the poststructuralist revolution, they believe in "human nature." That is, they believe that humans in all periods and cultures display a common, basic set of motives, feelings, and ways of thinking (Brown 1991). The literary Darwinists look to evolutionary social science to provide the most thorough, detailed guide to the actual content and structure of human nature, and they use that guide in analyzing the content and form of literary depictions, the perspectives of authors, and the responses of readers.

Quantifying Literary Meaning

The advent of a new critical vocabulary naturally initiates a phase of redescriptive commentary on the standard body of canonical literary texts. Evolutionary literary critics have already had considerable success in gaining a sharper focus on the themes that provide the skeletal structure for a number of specific literary works (for examples, see Scalise Sugiyama 2001b; Carroll 2004, 129–145, 163–185, 206–113; Nordlund 2007; Saunders 2007; Gottschall 2008; Boyd 2009; Saunders 2009; Winkelman 2009; Boyd et al. 2010; Clasen 2010; Duncan 2010; Swirski 2010; Carroll 2011b, 2012a, 2012b; Clasen 2012; Jonsson 2012; Saunders 2012; Carroll 2013a, 2013b; Clasen 2014; Carroll 2015). This process has only just begun, and there are thousands of occasions legitimately open for the process of redescription within a more adequate critical vocabulary. Moreover, the source theories of evolutionary literary criticism are empirical and progressive. Evolutionary literary criticism can progress in tandem with the advance of knowledge in evolutionary social science.

The process of redescription within a vocabulary congruent with empirically derived knowledge is an important phase of Darwinian literary study. It is nonetheless a kind of study that accepts the limitations inherent within all purely discursive, humanistic commentary. It respects the canons of empirical probability, but it cannot submit its findings to impartial tests or assess alternative hypotheses by making predictions and analyzing data statistically. Consequently, it cannot produce empirically grounded concepts of its own.

In the research described in this chapter, we seek to bridge the gap between humanistic literary criticism and the empirical methodology of the social sciences. We produce data that can constrain interpretive criticism and that can also be incorporated into future empirical studies that quantify literary meaning. We make use of concepts available within evolutionary psychology, and we aim to produce findings that can, in turn, contribute to the development of theory and research within evolutionary psychology.

Building on findings in the evolutionary human sciences, we constructed a model of human nature, incorporated the model in on online questionnaire, and used responses to the questionnaire to illuminate the evolved psychology that shapes the organization of characters in nineteenth-century British novels (Austen to Forster). We induced hundreds of readers to give numerical ratings to the attributes of hundreds of characters. Participants also rated their own emotional responses to the characters.

The questionnaire was designed to test one central hypothesis: the idea that "agonistic structure" shapes the organization of characters in these novels. Respondents identified characters as protagonists, antagonists, or minor characters. We delineate the features that distinguish these groups of characters, demonstrate that relations among the groups form a central organizing principle in the novels, and propose an explanation for the adaptive function of agonistic structure.

Agonistic structure has a wide conceptual scope in its own right, but in analyzing agonistic structure, we are also serving a deeper purpose. By constructing a research design that correlates the features of characters with the responses of readers, we seek to produce a first approximation to a universal set of categories for analyzing meaning structures in fictional narratives. We believe that literary meaning is a natural phenomenon. Like all other natural phenomena, it can be reduced to constituent parts, measured, and located precisely within the causal network of nature. This broad supposition stands in sharp contrast to a belief, common in the humanities, that literary meaning is illimitably complex and contains irreducible elements of the qualitatively unique (Goodheart 2007; Crews 2008; Deresiewicz 2009; Goodheart 2009; Smee 2009; Spolsky 2009). No one study could confirm definitively that all literary meaning can be analyzed objectively, but individual studies can provide strong evidence that major features of meaning can be reduced effectively to simple categories grounded in an evolutionary understanding of human nature. By quantifying literary meaning, we are translating a naturalistic interpretive vision into empirical evidence that literary meaning is determinate, delimited in scope, and consilient with the knowledge of evolutionary biology.

Collecting Data

Consulting websites for English literature faculty at universities on several continents, we identified scholars interested in 19th-century British literature, especially fiction. We sent e-mails inviting these scholars to go to the website, select one or more characters, and fill out a questionnaire on each character selected.

Similar invitations were posted on listservs dedicated to the literature of the period or to individual authors in the period.

Approximately 519 respondents completed a total of 1470 questionnaires on 435 characters from 134 novels. A copy of the questionnaire used in the study can be accessed at http://www-personal.umich.edu/~kruger/carroll-survey.html. (Please note that the form is no longer active and is not used to collect data.) Many characters were coded multiple times. For example, the most popular character, Elizabeth Bennet from Jane Austen's *Pride and Prejudice,* received 81 codings. A little more than half the characters (53%) received only one coding. For characters who received multiple codings, we averaged the results. The results for any one character were counted only once in the total set of scores.

The respondents provided demographic information indicating that 178 (34%) were male and 341 (66%) were female. The majority of the respondents ranged between 25 years and 55 years of age. Eighty-one percent of the respondents had a bachelor's degree or higher, 58% had advanced degrees, and 32% had doctorates. The quality of the data indicates the respondents were, on the whole, well informed and took their task seriously.

The scores on motives, the criteria for selecting mates, and emotional responses produced data that we condensed into smaller sets of categories through factor analysis. The five personality domains represent a condensation of traits from six decades of factor analytic studies (John and Srivastava 1999; Barenbaum and Winter 2008). In this chapter, further condensing the results, we compare only protagonists and antagonists, and we display the results only for motives, long-term mating, personality, and emotional responses. These results bring out the main tendencies in the data. (For details omitted here, see Carroll et al. [2012].)

The Research Design

The questionnaire was designed to reduce the components of human nature to discrete categories, reduce the categories to finite sets of elements, and simulate, in protocol form, the socially interactive situation of a fictional narrative: an author, who is a person, talking about characters, who are fabricated persons, describing those characters and their actions for readers, who are also persons, and eliciting emotional and evaluative responses to those characters. The causal flow in Figure 12.1 forms a feedback loop. The designs or intentions of the author, in the top left-hand corner of the diagram, are the starting point in a causal sequence. The end point in the sequence—the creation of normative values within the novels of a given culture—feeds back into the designs or intentions of the author. Authors determine a character's attributes such as sex, age, attractiveness, personality, motives, and preferences in marital partners.

Readers respond emotionally to characters, wish them to succeed or fail in achieving their goals, and recognize whether the character is a major or minor character. On the basis of those responses, readers decide whether the character is a protagonist, an antagonist, a good minor character (associate of a protagonist), or a bad minor character (associate of an antagonist). Protagonists and their associates embody the positive values that authors anticipate their readers will share with the authors. Antagonists and their associates elicit morally and emotionally negative evaluative responses. The array of positive and negative evaluations elicited from readers is the "ethos" of the novel. The common features in the ethos of multiple novels in a given period reflects the ethos of the culture as a whole. Readers form an imaginative community of shared experience in their responses to the novels. Authors are influenced by the ethos of the culture in which they live. They recognize the shared values of that culture and design characters who will elicit predictable emotional responses from their readers.

The four agonistic roles—protagonists, antagonists, good minor characters, and bad minor characters—were divided into male and female sets, thus producing eight character sets in total. Organizing characters into these eight sets forms an implicit empirical hypothesis: that agonistic structure, differentiated by sex, is a fundamental shaping feature in the organization of characters in the novels.

We predicted (a) that each of the eight character sets would be sharply defined by a distinct and integrated array of features, that these features would correlate in sharply defined ways with the emotional responses of readers, and that both

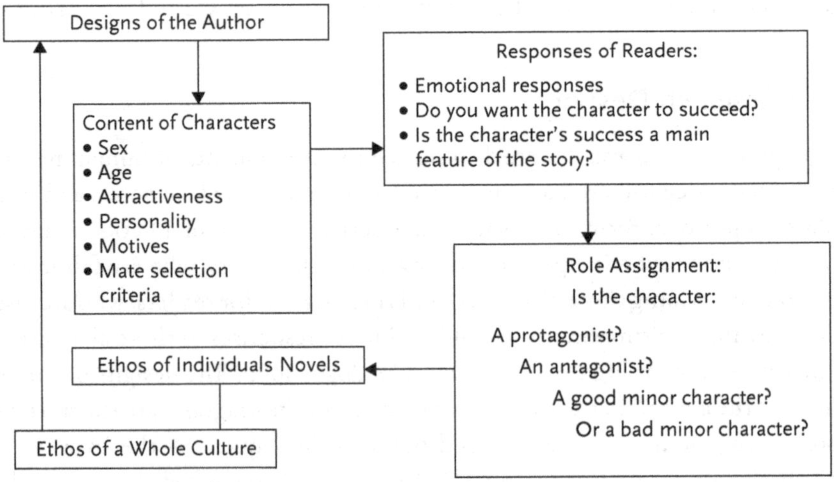

FIGURE 12.1 Research design.

the features of characters and the emotional responses of readers would correlate, on the average, with character role assignments; (b) that characters identified as protagonists, and their friends and associates, would have attributed to them, on average, the features to which readers are most attracted and most admire; (c) that characters identified as antagonists, and their friends and associates, would have attributed to them, on average, the characteristics for which readers feel an aversion and of which they disapprove; (d) that protagonists would realize most completely the approbatory tendencies in reader response; and (e) that antagonists would realize most completely the aversive tendencies.

Taken individually, each of these propositions might seem obvious, but only if one presupposes the validity of the terms *protagonist* and *antagonist*—the very terms our study was designed to test. To our knowledge, no previous empirical study has tested the validity of these terms. Their validity is not self-evident. One could argue reasonably enough that (a) novels reflect social and psychological reality, and that (b) in reality people are not divided into morally polarized groups. If this argument was correct, the characters in the novels would presumably display a complex and situationally contingent blend of morally valenced forms of behavior. One might then conclude that the terms *protagonist* and *antagonist* are bits of "folk" wisdom that fail to cut fictional narrative at its joints.

The four main categories on which we collected scores—motives, mating, personality, and emotional responses—should be able to give decisive evidence regarding whether agonistic structure forms a central structural principle in the novels. Motives are the basis for action in human life (McAdams 2009). Selecting a sexual or marital partner enters crucially into reproductive success and evokes, accordingly, exceptionally strong feelings (Buss 2000, 2003; Gottschall and Nordlund 2006). Personality traits are dispositions to act on motives (Nettle 2007; McAdams 2009). Emotions are the proximal mechanisms that activate motives and guide our social judgments, including our judgments of imaginary people (Feagin 1997; Plutchik 2003; McEwan 2005; Ekman 2007; Oatley et al. 2012).

These four categories take in a broad swath of human experience, the depiction of characters in novels, and readers' responses to those depictions. If the agonistic patterns produced by the categories had been vague in outline and inconsistent in their relations to one another, that result would have strongly suggested that agonistic structure does not account for much in the organization of characters in the novels. As it turns out, though, the patterns are not vague and inconsistent. They are clear and robust.

While testing for the validity of agonistic structure, we were also inquiring into the actual content of the attributes in characters that produce approbatory and aversive responses. Our most general supposition about those attributes was that protagonists would have prosocial dispositions and that antagonists would not. Implicit in that hypothesis was the idea that the novels form a medium through which authors and their readers affirm their membership within a community dependent on shared norms of cooperative behavior.

Results on Motives, Selecting Mates, Personality, and Emotional Responses

Motives

The most comprehensive scientific concepts for the systemic organization of the phases and functional roles of human life derive from "human life history theory." All species have a "life history," a species-typical pattern for birth, growth, reproduction, social relations (if the species is social), and death. For each species, the pattern of life history forms a reproductive cycle. In the case of humans, that cycle centers on parents, children, and the social group. Successful parental care produces children capable, when grown, of forming adult pair bonds, becoming functioning members of a community, and caring for children of their own. "Human nature" is the set of species-typical characteristics regulated by the human reproductive cycle (MacDonald 1997; Kaplan et al. 2009; Carroll 2011a; Muehlenbein and Flinn 2011).

For the purposes of this study, we divided human life history into a set of 12 basic motives—that is, goal-oriented behaviors regulated by the reproductive cycle. For survival, we included two motives—survival itself (fending off immediate threats to life) and performance of routine work to earn a living. We also asked about the importance of acquiring wealth, power, and prestige, and about the importance of acquiring a mate in both the short term and the long term. In the context of these novels, short-term mate selection would mean flirtation or illicit sexual activity; long-term mate selection would mean seeking a marital partner. Taking account of "reproduction" in its wider significance of replicating genes one shares with kin ("inclusive fitness"), we asked about the importance of helping offspring and other kin. For motives oriented to positive social relations beyond one's own kin, we included a question on "acquiring friends and making alliances" and another on "helping nonkin." And finally, to capture the uniquely human dispositions for acquiring complex forms of culture, we included "seeking education or culture" and "building, creating, or discovering something."

We predicted (a) that protagonists would be generally affiliative in their motives—concerned with helping kin and making friends; (b) that antagonists would be chiefly concerned with acquiring wealth, power, and prestige; and

(c) that protagonists would, on average, be much more concerned than antagonists or minor characters with acquiring education and cultural knowledge.

When we submitted scores on the 12 separate motives to factor analysis, five main factors emerged: social dominance, constructive effort, romance, subsistence, and nurture. Seeking wealth, power, and prestige all have strong positive loadings on social dominance; helping nonkin has a moderate negative loading. (That is, helping nonkin correlates negatively with seeking wealth, power, and prestige.) Constructive effort was defined most strongly by loadings from the two cultural motives, seeking education or culture, and creating, discovering, or building something' and also by loadings from two prosocial or affiliative motives: making friends and alliances and helping nonkin. Romance is a mating motive, chiefly loading on short-term and long-term mating. Subsistence combines two motives: survival and performance of routine tasks to gain a livelihood. Nurture is defined most heavily by loadings from nurturing/fostering offspring or other kin, and that motive correlates negatively with short-term mating. Helping nonkin also contributes moderately to this factor, bringing affiliative kin-related behavior into association with generally affiliative social behavior.

Male and female antagonists both display a pronounced and exclusive emphasis on social dominance (Figure 12.2).

Male protagonists score higher than any other character set on constructive effort and on subsistence. Female protagonists score higher than any other character set on romance, but their positive motives are fairly evenly balanced among constructive effort, romance, and nurture. In these novels, female protagonists are largely restricted to the nubile age range. That restriction corresponds with a pronounced emphasis on romance as a motive.

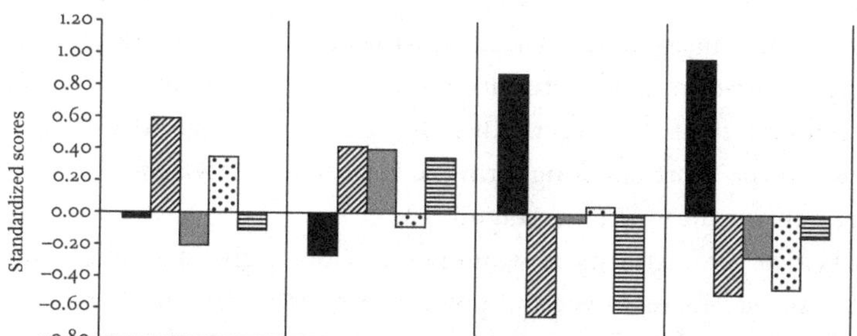

	Male protags	Female protags	Male antags	Female antags
Dominance	−0.03	−0.26	0.87	0.97
Constructive Effort	0.59	0.41	−0.64	−0.5
Romance	−0.2	0.4	−0.05	−0.26
Subsistence	0.34	−0.08	0.04	−0.46
Nurture	−0.11	0.34	−0.61	−0.14

FIGURE 12.2 Motive factors for protagonists and antagonists.

Criteria for Selecting Mates

Evolutionary psychologists have identified mating preferences that males and females share and also preferences that differ by sex. Males and females both value kindness, intelligence, and reliability in mates. Males preferentially value physical attractiveness, and females preferentially value wealth, prestige, and power. These sex-specific preferences are rooted in the logic of reproduction and have become part of human nature because they had adaptive value in ancestral environments. Physical attractiveness in females correlates with youth and health—hence, with reproductive potential. Wealth, power, and prestige enable a male to provide for a mate and her offspring (Buss 2003; Gangestad 2007; Geary 2010). We anticipated that scores for mate selection would correspond to the differences between males and females found in studies of mate selection in the real world. Because protagonists typically evoke admiration and liking in readers, we anticipated that protagonists would give stronger preference than antagonists to intelligence, kindness, and reliability. We reasoned that a preference for admirable qualities in a mate would evoke admiration in readers.

We asked questions about selecting mates in both the short term and the long term. In the results of the factor analyses for mate selection, the loadings for short-term and long-term mating are almost identical and divide with the sharpest possible clarity into three distinct factors: extrinsic attributes (a desire for wealth, power, and prestige in a mate), intrinsic qualities (a desire for kindness, reliability, and intelligence in a mate), and physical attractiveness (that one criterion by itself).

We anticipated differences in mate preferences in the short and long term, but our respondents evidently read the question on short-term mating to mean something different from what we had in mind. We had in mind illicit sexual activity. But respondents gave scores on short-term mating to many characters who do not engage in illicit sex. In many cases, the respondents evidently interpreted short-term mating to mean any romantic excitement in its early phases, even for relations that eventually culminate in marriage. The scores on selecting mates in the short and long term are essentially equivalent. We give the results here only for the long term (Figure 12.3).

Female protagonists and antagonists both give a stronger preference to extrinsic attributes—wealth, power, and prestige—than male protagonists or antagonists, but female antagonists exaggerate the female tendency toward preferring extrinsic attributes. The emphasis female antagonists give to extrinsic attributes parallels their single-minded pursuit of social dominance. Female protagonists give a more marked preference than male protagonists to intrinsic qualities—intelligence, kindness, and reliability.

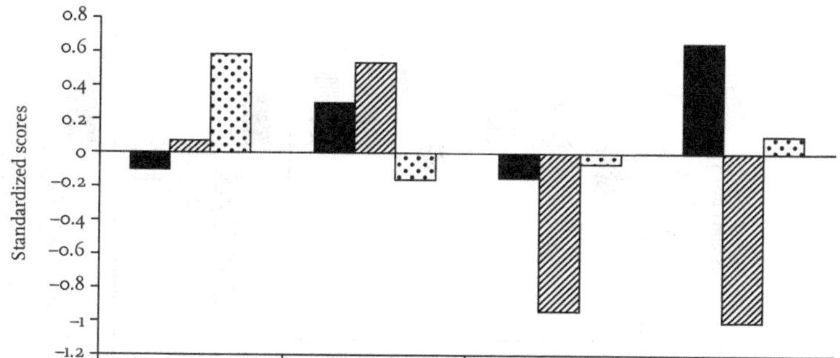

	Protagonists: Males	Protagonists: Females	Antagonists: Males	Antagonists: Females
Extrinsic	−0.1	0.29	−0.15	0.64
Intrinsic	0.07	0.53	−0.94	−1.01
Attractiveness	0.57	−0.16	−0.05	0.1

FIGURE 12.3 Mate selection criteria for protagonists and antagonists.

Personality Factors

The standard model for personality is the five-factor or "Big Five" model. *Extraversion* signals assertive, exuberant activity in the social world versus a tendency to be quiet, withdrawn, and disengaged. *Agreeableness* signals a pleasant, friendly disposition and tendency to cooperate and compromise versus a tendency to be self-centered and inconsiderate. *Conscientiousness* refers to an inclination toward purposeful planning, organization, persistence, and reliability versus impulsivity, aimlessness, laziness, and undependability. *Emotional stability* reflects a temperament that is calm and relatively free from negative feelings versus a temperament marked by extreme emotional reactivity and persistent anxiety, anger, or depression. *Openness to experience* describes a dimension of personality that distinguishes open (imaginative, intellectual, creative, complex) people from *closed* (down-to-earth, uncouth, conventional, simple) people (John et al. 1988; Johnson and Ostendorf 1993; Costa and McCrae 1997; Saucier and Ostendorf 1999; Nettle 2007; McAdams 2009).

We predicted that (a) protagonists and their friends would, on average, score higher on the personality factor of agreeableness, a measure of warmth and affiliation; and (b) that protagonists would score higher than antagonists on openness to experience, a measure of intellectual vivacity.

Male and female protagonists are both somewhat introverted, agreeable, conscientious, emotionally stable, and open to experience (Figure 12.4).

Female protagonists score higher than any other set on agreeableness, conscientiousness, and openness, and they score in the positive range on stability. In personality, male protagonists look like slightly muted or moderated versions

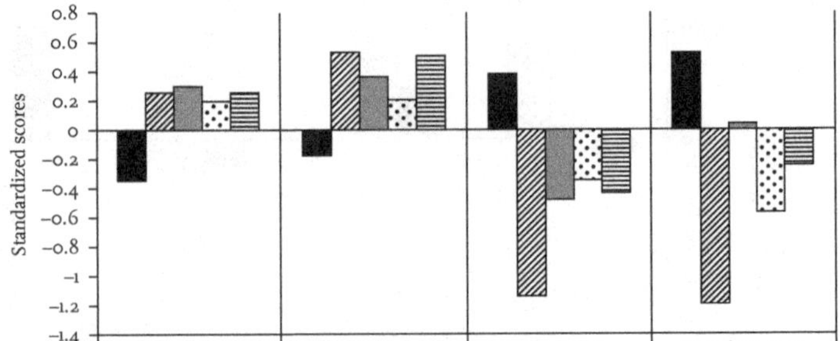

	Male protags	Female protags	Male antags	Female antags
Extraversion	−0.34	−0.17	0.38	0.52
Agreeableness	0.25	0.52	−1.13	−1.19
Conscientiousness	0.3	0.35	−0.47	−0.04
Stability	0.2	0.21	−0.34	−0.55
Openness	0.25	0.5	−0.42	−0.24

FIGURE 12.4 Personality traits of protagonists and antagonists.

of female protagonists. Male and female antagonists are both relatively extraverted, highly disagreeable, and low in stability and openness. On each of the five factors, the protagonists and antagonists pair off and stand in contrast to one another.

Emotional Responses

One of our chief working hypotheses is that when readers respond to characters in novels, they respond in much the same way, emotionally, as they respond to people in everyday life. They like or dislike them, admire them or despise them, fear them, feel sorry for them, or are amused by them. In writing fabricated accounts of human behavior, novelists select and organize their material for the purpose of generating such responses, and readers willingly cooperate with this purpose. They participate vicariously in the experiences depicted and form personal opinions about the qualities of the characters. Authors and readers thus collaborate in producing a simulated experience of emotionally responsive evaluative judgment (Bower and Morrow 1990; Grabes 2004; Mar and Oatley 2008; Oatley 2011).

We sought to identify emotions that are universal and that are thus likely to be grounded in universal, evolved features of human psychology. The solution was to use Paul Ekman's (2007) influential set of seven basic or universal emotions: anger, fear, disgust, contempt, sadness, joy, and surprise. These terms were adapted for the purpose of registering graded responses specifically to persons or characters. Four of the seven terms were used unaltered: anger, disgust, contempt, and sadness. Fear was divided into two distinct items: fear *of* a character and fear *for* a character. "Joy" or "enjoyment" was adapted both to make it idiomatically appropriate as a response to a person and also to have it register some distinct qualitative differences. Two terms, *liking* and *admiration*, served these purposes. "Surprise," like "joy," seems more appropriate as a descriptor for a response to a situation than as a descriptor for a response to a person or character. Consequently, in place of the word *surprise*, we used the word *amusement*, which combines the idea of surprise with an idea of positive emotion. One further term was included in the list of possible emotional responses: indifference. Indifference is the flip side of "interest," the otherwise undifferentiated sense that something matters, that it is important and worthy of attention.

We predicted (a) that protagonists would receive high scores on the positive emotional responses liking and admiration; (b) that antagonists would receive high scores on the negative emotions anger, disgust, contempt, and fear *of* the character; (c) that protagonists would score higher on sadness and fear *for* the character than antagonists; and (d) that major characters (protagonists and antagonists) would score lower on indifference than minor characters.

Factor analysis produced three clearly defined emotional response factors: (a) dislike, which includes anger, disgust, contempt, and fear *of* the character, and which also includes negative correlations with admiration and liking; (b) sorrow, which includes sadness and fear *for* the character and a negative correlation with amusement; and (c) interest, which consists chiefly of a negative correlation with indifference.

Male and female protagonists both scored relatively low on dislike and relatively high on sorrow (Figure 12.5). Male and female antagonists scored very high on dislike—higher than any other set—low on sorrow, and somewhat above average on interest. Female protagonists scored high on interest, but male

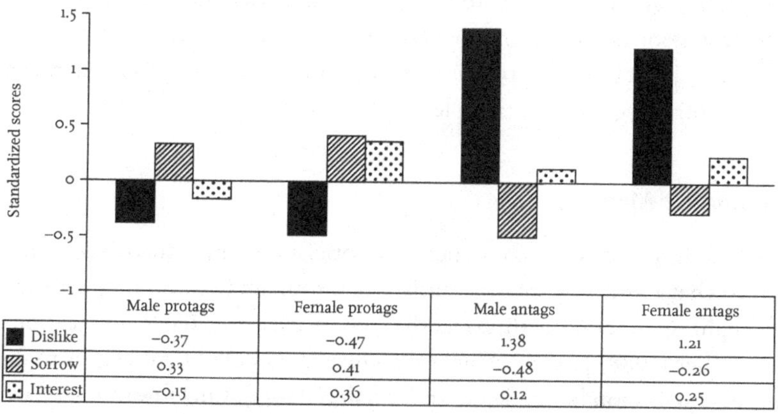

	Male protags	Female protags	Male antags	Female antags
Dislike	−0.37	−0.47	1.38	1.21
Sorrow	0.33	0.41	−0.48	−0.26
Interest	−0.15	0.36	0.12	0.25

FIGURE 12.5 Emotional responses to protagonists and antagonists.

protagonists scored below average on interest. They scored lower even than good minor males, although not lower than the other minor characters.

The relatively low score received by male protagonists on interest ran contrary to our expectation that protagonists, both male and female, would score lower on indifference than any other character set. We explain this finding by appealing to the psychology of cooperation. Male protagonists in our data set are introverted and agreeable, and they do not seek to dominate others socially. They are mild, pleasant, reliable, and intellectually curious. They are not very assertive and thus do not excite much competitive antagonism. They exemplify normative values of cooperative behavior. By integrating themselves so thoroughly into the group, suppressing or masking their desire for dominance, they diminish the level of attention that other members of the group give to them.

Conclusion

Agonistic Polarization

The characters in these novels display an integrated array of agonistically polarized attributes, and readers respond to those attributes in emotionally polarized ways. The antagonists are preoccupied with wealth, prestige, and power—egoistic striving wholly segregated from social affiliations. Male antagonists are indifferent to the personal qualities in their marital partners. Female antagonists choose partners solely on the basis of wealth and status. Antagonists are both emotionally isolated and also incurious. Protagonists cultivate friendships, seek romantic love, and pursue cultural interests. They are emotionally warm, conscientious, and broad-minded. The polarized emotional responses of readers correlate strongly with this integrated array of attributes. Readers respond with aversion and disapproval to antagonists and with admiration and sympathy to protagonists. We can reasonably conclude that, in these novels, agonistic structure—the polarized opposition of moral and personal traits in characters—forms a central organizing principle.

Determinate Meaning

Under the influence of deconstructive skepticism, literary theorists have often affirmed that meanings are inherently indeterminate because they are inescapably caught up in semiotic slippages that produce irreconcilable implications. As D. A. Miller (1988) put it: "Whenever a text makes confident claims to cognition, these will soon be rendered undecidable" (x–xi). Our findings lead us to a different conclusion.

Psychologists presuppose that when multiple respondents agree about features of people, those features actually exist. The subjects in this study are imagined people rather than actual people, but the principle is the same. We found high levels of agreement on the attributes of characters and thus assume those attributes actually exist in the characters. Moreover, we found a high degree of correlation between attributed features and the emotional responses of readers. If the features readers identify in characters actually exist, those features are determined by authors. Authors stipulate a character's sex, age, personality, motives, and criteria for selecting mates. They also stipulate the character's actions, which are based on motives and the personality dispositions that orient characters to motives. Our data indicate that readers largely agree in recognizing and identifying character attributes. If readers' emotional responses to characters correlate strongly with the characters' attributes, and if readers tend to respond in emotionally similar ways to those attributes, we can reasonably infer that authors have a high degree of control in determining readers' emotional responses.

Can any one set of descriptive and analytic terms (or their close synonyms) be used to define meaning in a given text; be assigned priority over other, competing terms; and be presented in such a way that the weight of empirical evidence confirms its validity overwhelmingly? The suppositions in most current literary theories tend to run strongly in the other direction. Case books on literary texts now typically contain essays representing various theoretical schools, most often deconstruction, psychoanalysis, feminism, and Marxism and/or New Historicism. That kind of pluralism suggests an underlying epistemological relativism. Using strong versions of Kuhn's theory of "paradigms," literary theorists have often affirmed that every structure of meaning changes systemically in accordance with the interpretive framework being used. In the most extreme version of this idea, meaning is always determined preemptively—essentially created—by an "interpretive community" (Fish 1980).

In our view, pluralism is not a coherent theoretical position. It is a rationalization of the epistemic disorder that characterizes most disciplines in the period before a paradigm has formed. A paradigm in science is a structure of theories that can provide the basis for a reasoned consensus among researchers committed to scientific criteria of epistemic validity. Reasoned consensus arises when a structure of theories is broad in scope, logically coherent, and concordant with empirical evidence. Paradigms are frameworks for continuing programs of cumulative empirical research. In our view, the evolutionary human sciences now provide an adequate basis for rational consensus. The evolutionary understanding of the adapted mind has clearly become a framework for a continuing program of empirical research.

Each of the theoretical schools currently active in literary criticism consists of propositions about matters susceptible to empirical confirmation or falsification: the nature of the human mind, the laws of social organization, language, sex, and gender. Some of these propositions are true, some partly true, and some false. The evolutionary understanding of human nature offers a vantage point from which to make reasoned assessments about what is true and false in the claims of the speculative theoretical schools. The evolutionary human sciences are grounded in evolutionary biology, itself firmly established as a scientific paradigm concordant with knowledge in the other sciences. Evolutionary biology has a prima facie epistemic validity that can be claimed by none of the competing theoretical schools in literary study. Whatever is empirically sound and conceptually rich in the speculations of other theoretical schools can be assimilated to the evolutionary paradigm.

For the first time in intellectual history, we have the basis for a rational interpretive consensus about the meaning of literary works. Within the framework of ideas provided by the evolutionary human sciences, meaning can become determinate from the lowest to the highest level of cognitive response: from the perceptual level to the level of analytic summary to the level at which the largest implications of these two lower levels are explained and linked with concepts across the whole field of the human sciences. (On these three levels of cognitive response, see Bordwell [2008, 43–53].)

We are not, of course, claiming that all writers are evolutionists. Historically, that is not even possible. We claim only that writers use the common idiom, that the common idiom contains determinate meanings, that all determinate meanings can ultimately be explained within the framework of evolutionary biology, and that evolutionary biology is a scientifically valid framework that encompasses and either subsumes or supplants all other competing theoretical systems in the human sciences.

Sexual Politics in the Novels

For several decades now, no feature in personal and social identity has received more critical attention than sex and gender. Much of this criticism has taken as its central theme struggles for power based on sex. Our data indicate that struggles for power based on sex are less important than the conflict between dominance and cooperation. Despite differences of sex, male and female protagonists are much more similar to each other than either are to male or female antagonists. Male and female antagonists, also, are much more similar to each other than either are to male or female protagonists. In the features that distinguish characters, being a protagonist or antagonist matters more than being male or female.

Our data suggest that in these novels conflict between the sexes is subordinated to their shared and complementary interests. In the agonistic structure of plot and theme, male and female protagonists are allies. They cooperate in resisting the predatory threats of antagonists, and they join together to exemplify the values that elicit readers' admiration and sympathy. Both male and female antagonists are massively preoccupied with material gain and social rank. That preoccupation stands in stark contrast to the more balanced and developed world of the protagonists—a world that includes sexual interest, romance, the care of family and friends, and the life of the mind. By isolating and stigmatizing dominance behavior, the novels affirm the shared values that bind its members into a community.

Agonistic structure appears in many imaginative contexts. It shapes any ideology in which a protagonistic group characterized by prosocial dispositions is set in sharp contrast to an alien group that personifies a will to domination. One such ideological context appears in feminist commentaries on British novels of the 19th century. Feminist criticism characteristically displays an agonistically polarized vision of human sexual identity. Within that vision, "patriarchy," the system of male domination, embodies the desire for social dominance as an end in itself. Patriarchy is thus a paradigmatically antagonistic force. The counterposing protagonistic force consists of a specifically female ethos of affiliative social interaction.

The agonistic structure of feminist theory incorporates two distinct concepts of human sexual identity. These two concepts are not logically consistent with one another, but they serve complementary imaginative functions. One concept is that "gender" consists exclusively of roles imposed arbitrarily by society and culture. The other concept is that males and females are radically separate forms of life characterized by independent and incompatible systems of affect, cognition, and value. The first concept is "constructivist," and the second, "essentialist" (Martin 1994; Dietz 2003; Schore 2003; Vandermassen 2005; Gaard 2011). In the constructivist concept of sexual identity, the anatomical, physiological, and neurological differences between males and females, if they are not actually produced by culture, consist of merely physical features that have no significant impact on motives, emotions, or behavior. In the essentialist concept, male and female dispositions are themselves primary and irreducible constituents of the moral universe. The constructivist concept of sexual identity dislodges sexual identity from the causal constraints of human life history, and the essentialist concept identifies human sex differences as autonomous moral forces within an agonistically polarized field of action.

Both the cultural constructivist and the essentialist conceptions of human sexual identity make contact with important aspects of human sexual reality.

Sociosexual roles vary greatly from culture to culture, but each sex also has genetically transmitted dispositions that transcend cultural differences and constrain cultural formations. The longer, Darwinian perspective on human life history captures the elements of truth in both these observations and integrates those elements into a comprehensive and consistent understanding of human sexual identity. The complex functional structures that distinguish males and females at the present time are features that, on average, contributed most effectively to the reproductive success of their ancestors. The sexes are not separate and autonomous systems of motivation and affect. Males and females have co-evolved, in reciprocally causal ways under the constraining force of partially shared and partially conflicting reproductive interests. Human males and females are reproductively interdependent. Human sexual relations require humans to negotiate conflicts between reciprocal benefits and competing interests, and in that respect, human sexual relations are like all other affiliative human social relations, including those of parents and children.

Human life history entails species-typical differences in mating preferences for males and females. Those differences are reflected in the preferences in the males and females in our data set. Plots are based on motives and desires; sex bulks large among the motives that drive plots, and biologically based mating preferences infuse passion and interest into motives. These observations have important implications for the interpretation of sex and gender in the novels. A critic who registers the way evolved sex differences shape stories and infuse them with passion and interest is unlikely to speak of the novels in quite the same way as a critic who believes that sex roles are determined solely by social convention. So also, in discussing the sexual politics in novels, a basic difference opens up between critics who see sex as a powerful, primary force and critics who see it chiefly as a medium for the circulation of sociopolitical energy. In an obvious way, all sex is political. That is, all sex is bound up with social power relations. But sex is not merely a product of social power relations. No cultural or literary theory that overlooks the deep adaptive history of human mating preferences is likely to capture the real force of sexual passion in novels, and without getting the sexual passion right, one cannot get the politics right either.

Sex is interesting in itself, but we find a still deeper interest in the interaction between sex and agonistic status. Female characters prefer extrinsic attributes in their mates, and male characters prefer physical attractiveness. Well and good—just what, from an evolutionary perspective, one would expect. But that is not the whole story, or even the main story. The main story concerns the opposition between good and bad characters, both male and female. Among the good characters, esteem and gratitude count for something, and romantic love is possible. The bad characters are interested in neither love nor sex. They are interested only in power, wealth, and prestige. In discussing sex and sexual politics in the novels, we need to be sure we are talking about the whole emotional world of the novels, not just about the preoccupations with power that distinguish the antagonists.

The Adaptive Function of Agonistic Structure

One of the most hotly debated issues in evolutionary studies in the humanities is whether the arts fulfill any adaptive function at all (Boyd 2005; Carroll 2008a, 119–128, 2008b, 349–368). Various theorists have proposed possible adaptive functions, for instance, reinforcing the sense of a common social identity (Dissanayake 2000; Boyd 2009), fostering creativity and cognitive flexibility (Boyd 2009), enhancing pattern recognition (Boyd 2009), serving as a form of sexual display (Miller 2000; Dutton 2009), providing information about the environment (Scalise Sugiyama 2001a), offering game-plan scenarios to prepare for future problem solving (Scalise Sugiyama 2005; Swirski 2006), focusing the mind on adaptively relevant problems (Dissanayake 2000; Tooby and Cosmides 2001; Salmon and Symons 2004), and making emotional sense of experience (Wilson 1998; Dissanayake 2000; Carroll 2008a, 2008b; Dutton 2009; Carroll 2012c).

One chief alternative to the idea that the arts provide *some* adaptive function is that literature and the other arts are like the color of muscle tissue or the sound produced by sneezing—a functionless side effect of adaptive processes (Pinker 1997, 2007). The data on agonistic structure point to a different conclusion. The ethos reflected in the agonistic structure of the novels replicates the egalitarian ethos of hunter-gatherers, who stigmatize and suppress status seeking in potentially dominant individuals. By supporting group solidarity, the egalitarian ethos fulfills an adaptive function for hunter-gatherers. If agonistic structure in the novels engages the same social dispositions that animate hunter-gatherers, our study would lend support to the hypothesis that literature can fulfill at least one adaptive social function.

In *Hierarchy in the Forest: The Evolution of Egalitarian Behavior*, Christopher Boehm (1999) offers a cogent explanation for the way interacting impulses of dominance and affiliation have shaped the evolution of human political behavior. During an earlier phase of the evolutionary human sciences, sociobiological theorists had repudiated the idea of "altruistic" behavior and had restricted prosocial dispositions to nepotism and to the exchange of reciprocal benefits. In contrast, Boehm (1999) argues that at some point in their evolutionary history—at the latest, 100,000 years ago—humans developed a special capacity, dependent on their symbolic and cultural capabilities, for enforcing altruistic or group-oriented norms. By enforcing these norms, humans succeed in controlling "free riders" or "cheaters," and they thus make it possible for genuinely altruistic genes to survive within a social group. The selection for altruistic dispositions—and dispositions for enforcing altruistic cultural norms—would involve a process of gene-culture co-evolution that would snowball in its effect of altering human nature itself.

We can reason backward from our findings on agonistic structure to formulate hypotheses about functions fictional narratives might have fulfilled in ancestral environments. By identifying one of the ways novels actually work for us now, we can produce evidence relevant to hypotheses about the evolutionary origin and adaptive function of the arts. Agonistic structure is a central principle in the organization of characters in the novels. Taking into account not just the representation of characters, but also the emotional responses of readers, we can identify agonistic structure as a simulated experience of emotionally responsive social interaction. That experience has a clearly defined moral dimension. Agonistic structure precisely mirrors the kind of egalitarian social dynamic documented by Boehm (1999) in hunter-gatherers—our closest contemporary proxy to ancestral humans. As Boehm (1999, 2012) and others (Wilson and Wilson 2007; Gintis and Van Schaik 2012; Haidt 2012; Richerson and Henrich 2012) have argued, the dispositions that produce an egalitarian social dynamic are deeply embedded in the evolved and adapted character of human nature. Humans have an innate desire for power and an innate dislike of being dominated. Egalitarianism as a political strategy arises as a compromise between the desire to dominate and the dislike of being dominated. By pooling their power to exercise collective social coercion, individuals in groups can repress dominance behavior in other individuals. The result is autonomy for individuals. No one gets all the power he or she would like, but then, no one has to accept submission to other dominant individuals. Boehm (1999) describes in detail the pervasive collective tactics for repressing dominance within social groups organized at the levels of bands and tribes.

An egalitarian social dynamic is the most important basic structural feature that distinguishes human social organization from the social organization of chimpanzees. In chimpanzee society, social organization is regulated exclusively by dominance—that is, power. In human society, social organization is regulated by interactions between impulses of dominance and impulses for suppressing dominance. State societies with elaborate systems of hierarchy emerged only very recently in the evolutionary past, about 6000 years ago, after the agricultural revolution made possible concentrations of resources and, therefore, power. Before the advent of despotism, the egalitarian disposition for suppressing dominance had, at a minimum, 100,000 years in which to become entrenched in human nature—more than sufficient time for significant adaptive change to take place (Mithen 1996; Klein 2002; Wade 2006; Mellars 2007; Cochran and Harpending 2009).

In highly stratified societies, dominance assumes a new ascendancy, but no human society dispenses with the need for communitarian association. It seems likely, then, that morally polarized forms of agonistic structure in fictional narratives emerged in tandem with specifically human adaptations for cooperation and specifically human adaptations for creating imaginative constructs that embody the ethos of the tribe.

Agonistic structure in these novels seems to serve as a medium for readers to participate vicariously in an egalitarian social ethos. If that is the case, the novels can be described as prosthetic extensions of social interactions that in nonliterate cultures require face-to-face interaction. If suppressing dominance in face-to-face interaction fulfills an adaptive function, and if agonistic structure in narrative is a cultural technology that extends that interaction by imaginative means, one could reasonably conclude that agonistic structure fulfills an adaptive function.

The Scope of Our Claims

On the basis of the data on this particular set of novels, we have drawn conclusions about the determinacy of meaning, sexual politics, and adaptive function. How far can we generalize from these conclusions to all literature? In every period and every culture? Logically, it is possible that no other literary texts anywhere in the world contain determinate meanings, display differences between protagonists and antagonists more prominent than differences between male and female characters, or fulfill any adaptive function at all. Hypothetically possible, but not very likely. If our arguments hold good for this body of texts, they demonstrate that determinate meaning is at least possible, that in at least one body of classic narratives, agonistic role assignment—being a protagonist or antagonist—looms larger than gender role assignment, and that the organization of characters in at least one important body of fictional narratives reflects evolved social dispositions that in ancestral populations fulfilled adaptive functions. It seems unlikely that in these three important respects this body of novels is wholly anomalous.

In proposing that agonistic structure in these novels fulfills an adaptive social function, we do not imagine we have isolated the sole adaptive function of all literature. Quite the contrary. Along with other evolutionary theorists, we strongly suspect that literature and its oral antecedents fulfill other functions. Even if it is only one among other possible adaptive functions for narrative and drama, could we reasonably conclude that agonistic structure is a human universal—a formal structure that would appear in the narrative and dramatic productions in all cultures, at all periods, everywhere in the world? We have argued that the social dynamics animating these novels derive from ancient, basic features of human nature. Such features would, in all likelihood, appear in some fictional narratives in most or all cultures. If morally polarized agonistic structure is in fact a human universal, we would be interested to know how it varies in form in different cultural ecologies. Marriage—the "publicly recognized right of sexual access to a woman deemed eligible for childbearing"—is a human universal, but varies in form from culture to culture (Brown 1991, 136). We might expect agonistic structure, like marriage, to vary in form. These questions would make good topics of research for other studies. Until those studies are conducted, though, the topics are only a matter for theoretical speculation. For this current study, we can positively affirm only the conclusions we think our data allow us to draw.

Limitations in Our Analytic Model

In its most complete forms, Darwinist literary criticism would construct continuous explanatory sequences linking "inclusive fitness"—the "ultimate" causal principle in evolution—to particular features in an evolved and adapted human nature and to particular structures and effects in specific works of art. A comprehensively adequate interpretive account of a given work of art would take in, synoptically, its phenomenal effects (tone, style, theme, formal organization); locate it in a cultural context; explain that cultural context as a particular organization of the elements of human nature within a specific set of environmental conditions (including cultural traditions); identify an implied author and an implied reader; examine the responses of actual readers (for instance, other literary critics); describe the sociocultural, political, and psychological functions the work fulfills; locate those functions in relation to the evolved needs of human nature; and link the work comparatively with other artistic works using a taxonomy of themes, formal elements, affective elements, and functions derived from a comprehensive model of human nature.

In the current study, how far have we succeeded in approximating to this ideal of a complete critical account of the texts we discuss? We can identify specific areas in which we fall short of it. We did not aim at a universal, exhaustive explanation of the novels. We focused on only one specific large-scale element in the organization of characters: agonistic structure differentiated by sex. We did not construct a complete taxonomy of formal elements. More particularly, we did not incorporate ways of operationalizing some of the concepts that form the subject matter of narratology—for instance, the distinction between *syuzhet* and *fabula* or distinctions among different types of narrators. Insofar as we are concerned with quantifying features in individual texts, the main gap in our research design is probably the absence of any means for registering verbal "style": diction, syntax, rhythm, metaphors, motifs, and figures of speech.

In practical reality, there are limitations to what can be done with any given protocol. At least one of us (Gottschall) concedes that certain kinds of literary problems might never be fully amenable to a quantitative methodology. At least one other of us (Carroll) believes that all mental phenomena, including those involved in the production and reception of novels, consist of states of the brain and are hypothetically susceptible to quantification. But here we enter the realm of science fiction—a genre that deliberately erases the boundaries between "reality" and what is only "hypothetically" possible.

If any such science fiction scenario could be realized, it still would not render the personal, subjective aspect of literary study obsolete. "Meaning" and "effect" are crucial elements of literary phenomenology, and meaning is always meaning *for* someone, some particular person; effect is always an effect *on* some particular person. Literary scholars explain their subjects, or try to, but they also register the value and significance of their subjects. Identifying large-scale patterns of meaning in the novels need not reduce our appreciation of the value and significance of the novels. Quite the contrary. The better we understand how the novels work, the more keenly we can appreciate their effects. True enough, when scholars succeed in narrowing the range of possibly valid conclusions, they reduce the sense of vaguely infinite potential in the world of literary response, but they also open up new possibilities for actual discovery—for deeper levels of explanation, more complete understanding.

References

Abrams, Meyer H. 1997. The transformation of English studies: 1930–1995. *Dædalus: Journal of the American Academy of Arts and Sciences* 126: 105–131.

Barenbaum, Nicole B., and David G. Winter. 2008. History of modern personality theory and research. In Oliver P. John, Richard W. Robins, and Lawrence A. Pervin, eds., *Handbook of personality: Theory and research*, 3rd ed., 3–28. New York: Guilford.

Boehm, Christopher. 1999. *Hierarchy in the forest: The evolution of egalitarian behavior.* Cambridge, MA: Harvard University Press.

———. 2012. *Moral origins: The evolution of virtue, altruism, and shame.* New York: Basic Books.

Boghossian, Paul. 2006. *Fear of knowledge: Against relativism and constructivism.* Oxford: Oxford University Press.

Bordwell, David. 2008. *Poetics of cinema.* New York: Routledge.

Bower, Gordon H., and Daniel G. Morrow. 1990. Mental models in narrative comprehension. *Science* 247: 44–48.

Boyd, Brian. 2005. Evolutionary theories of art. In Jonathan Gottschall and David Sloan Wilson, eds., *The literary animal: Evolution and the nature of narrative*, 147–176. Evanston, IL: Northwestern University Press.

———. 2009. *On the origin of stories: Evolution, cognition, and fiction*. Cambridge, MA: Harvard University Press.

Boyd, Brian, Joseph Carroll, and Jonathan Gottschall, eds. 2010. *Evolution, literature, and film: A reader*. New York: Columbia University Press.

Brown, Donald E. 1991. *Human universals*. Philadelphia: Temple University Press.

Buss, David M. 2000. *The dangerous passion: Why jealousy is as necessary as love and sex*. New York: Free Press.

———. 2003. *The evolution of desire: Strategies of human mating*. New York: Basic Books.

Carroll, Joseph. 2004. *Literary Darwinism: Evolution, human nature, and literature*. New York: Routledge.

———. 2008a. An evolutionary paradigm for literary study. *Style* 42: 103–135.

———. 2008b. Rejoinder to the responses. *Style* 42: 308–411.

———. 2011a. Human life history and gene-culture co-evolution: An emerging paradigm. *Evolutionary Review: Art, Science, Culture* 2: 23–37.

———. 2011b. *Reading human nature: Literary Darwinism in theory and practice*. Albany: State University of New York Press.

———. 2012a. An evolutionary approach to *King Lear*. In John Knapp, ed., *Critical insights: The family*, 83–103. Ipswich, MA: EBSCO.

———. 2012b. Meaning and effect in fiction: An evolutionary model of interpretation illustrated with a reading of "Occurrence at Owl Creek Bridge." *Style* 26: 297–316.

———. 2012c. The truth about fiction: Biological reality and imaginary lives. *Style* 46: 129–160.

———. 2013a. Correcting for *The Corrections*: A Darwinian critique of a Foucauldian novel. *Style* 47: 87–118.

———. 2013b. Violence in literature: An evolutionary perspective. In Todd K. Shackelford and Ranald D. Hansen, eds., *Evolution of violence*, 33–52. New York: Springer.

———. 2015. Evolved human sociality and literature. In Jonathan H. Turner, Richard Machalek, and Alexandra Maryanski, eds., *Handbook on evolution and society: Toward an evolutionary social science*, 572–608. Boulder, CO: Paradigm.

Carroll, Joseph, Jonathan Gottschall, John A. Johnson, and Daniel Kruger. 2012. *Graphing Jane Austen: The evolutionary basis of literary meaning*. New York: Palgrave Macmillan.

Clasen, Mathias F. 2010. Vampire apocalypse: A biocultural critique of Richard Matheson's *I am legend*. *Philosophy and Literature* 34: 313–328.

———. 2012. Attention, predation, counterintuition: Why Dracula won't die. *Style* 43: 396–416.

———. 2014. Evil monsters in horror fiction: An evolutionary perspective on form and function. In Sharon Packer and Jody Pennington, eds., *A History of Evil in Popular Culture: What Hannibal Lecter, Stephen King, and Vampires Reveal About America*, vol. 2, 39–47. Santa Barbara, California: ABC-CLIO/Praeger.

Cochran, Gregory, and Henry Harpending. 2009. *The 10,000 year explosion: How civilization accelerated human evolution.* New York: Basic Books.

Costa, Paul T., Jr., and Robert R. McCrae. 1997. Personality trait structure as a human universal. *American Psychologist* 52: 509–516.

Crews, Frederick. 2008. Apriorism for empiricists. *Style* 42: 155–160.

Deresiewicz, William. 2009. Adaptation: On literary Darwinism. *Nation,* June 8, 26–31.

Dietz, Mary G. 2003. Current controversies in feminist theory. *Annual Review of Political Science* 6: 399–431.

Dissanayake, Ellen. 2000. *Art and intimacy: How the arts began.* Seattle: University of Washington Press.

Duncan, Charles. 2010. Darkly Darwinian parables: Ian McEwan and *The Comfort of Strangers*. *Evolutionary Review: Art, Science, Culture* 1: 120–124.

Dutton, Denis. 2009. *The art instinct: Beauty, pleasure, and human evolution.* New York: Bloomsbury Press.

Ekman, Paul. 2007. *Emotions revealed: Recognizing faces and feelings to improve communication and emotional life.* New York: Owl Books.

Feagin, Susan L. 1997. Imagining emotions and appreciating fiction. In Mette Hjort and Sue Laver, eds., *Emotion and the arts,* 50–62. New York: Oxford University Press.

Fish, Stanley Eugene. 1980. *Is there a text in this class? The authority of interpretive communities.* Cambridge, MA: Harvard University Press.

Fromm, Harold. 1997. My science wars. *The Hudson Review* 49: 599–609.

Gaard, Greta. 2011. Ecofeminism revisited: Rejecting essentialism and re-placing species in a material feminist environmentalism. *Feminist Formations* 23: 26–53.

Gangestad, Steven W. 2007. Reproductive strategies and tactics. In Robin I. M. Dunbar and Louise Barrett, eds., *Oxford handbook of evolutionary psychology*, 321–332. Oxford: Oxford University Press.

Geary, David C. 2010. *Male, female: The evolution of human sex differences.* Washington, DC: American Psychological Association.

Gintis, Herbert, and Carel P. Van Schaik. 2012. Zoon politicon: The evolutionary roots of human sociopolitical systems. In Peter J. Richerson and Morten H. Christiansen, eds., *Cultural evolution*, 25–44. Cambridge, MA: MIT Press.

Goodheart, Eugene. 2007. *Darwinian misadventures in the humanities.* New Brunswick, NJ: Transaction.

———. 2009. Do we need literary Darwinism? *Style* 42: 181–185.

Gottschall, Jonathan. 2008. *The rape of Troy: Evolution, violence, and the world of Homer.* Cambridge: Cambridge University Press.

Gottschall, Jonathan, and Marcus Nordlund. 2006. Romantic love: A literary universal? *Philosophy and Literature* 30: 450–470.

Grabes, Herbert. 2004. Turning words on the page into "real" people. *Style* 38: 221–235.

Graff, Gerald. 2007. *Professing literature: An institutional history*. Chicago: University of Chicago Press.

Gross, Paul R., and N. Levitt. 1994. *Higher superstition: The academic left and its quarrels with science*. Baltimore, MD: Johns Hopkins University Press.

Gross, Paul R., Norman Levitt, and Martin W. Lewis, eds. 1996. *The flight from science and reason*. New York: The New York Academy of Sciences.

Haidt, Jonathan. 2012. *The righteous mind: Why good people are divided by politics and religion*. New York: Pantheon Books.

John, Oliver P., Alois Angleitner, and Fritz Ostendorf. 1988. The lexical approach to personality: A historical review of trait taxonomic research. *European Journal of Personality* 2: 171–203.

John, Oliver P., and Sanjay Srivastava. 1999. The Big Five trait taxonomy: History, measurement, and theoretical perspectives. In Lawrence A. Pervin and Oliver P. John, eds., *Handbook of personality: Theory and research*, 2nd ed., 102–138. New York: Guilford.

Johnson, John A., and Fritz Ostendorf. 1993. Clarification of the five-factor model with the Abridged Big Five Dimensional Circumplex. *Journal of Personality and Social Psychology* 65: 563–576.

Jonsson, Emelie. 2012. "Man is the measure": Forster's evolutionary conundrum. *Style* 46: 161–176.

Kaplan, Hillard, Michael Gurven, and Jeffrey Winking. 2009. An evolutionary theory of human life span: Embodied capital and the human adaptive complex. In Vern L. Bengston, Daphna Gans, Norella M. Pulney, and Merril Silverstein, eds., *Handbook of theories of aging*, 2nd ed., 39–60. New York: Springer.

Klein, Richard G., with Blake Edgar. 2002. *The dawn of human culture*. New York: Wiley.

Koertge, Noretta. 1998. *A house built on sand: Exposing postmodernist myths about science*. New York: Oxford University Press.

MacDonald, Kevin 1997. Life history theory and human reproductive behavior. *Human Nature* 8: 327–359.

Mar, Raymond A., and Keith Oatley. 2008. The function of fiction is the abstraction and simulation of social experience. *Perspectives on Psychological Science* 3: 173–192.

Martin, Jane Roland. 1994. Methodological essentialism, false difference, and other dangerous traps. *Signs* 19: 630–657.

McAdams, Dan P. 2009. *The person: An introduction to the science of personality psychology*. New York: Wiley.

McEwan, Ian. 2005. Literature, science, and human nature. In Jonathan Gottschall and David Sloan Wilson, eds., *The literary animal: Evolution and the nature of narrative*, 5–19. Evanston, IL: Northwestern University Press.

Mellars, Paul, ed. 2007. *Rethinking the human revolution: New behavioural and biological perspectives on the origin and dispersal of modern humans.* Cambridge, UK: McDonald Institute for Archaeological Research.

Miller, D. A. 1988. *The novel and the police.* Berkeley: University of California Press.

Miller, Geoffrey F. 2000. *The mating mind: How sexual choice shaped the evolution of human nature.* New York: Doubleday.

Mithen, Steven J. 1996. *The prehistory of the mind: A search for the origins of art, religion, and science.* London: Thames and Hudson.

Muehlenbein, Michael P., and Mark V. Flinn. 2011. Patterns and processes of human life history evolution. In Thomas Flatt and Andreas Heyland, eds., *Mechanisms of life history evolution: The genetics and physiology of life history traits and trade-offs*, 153–168. Oxford: Oxford University Press.

Nettle, Daniel. 2007. *Personality: What makes you the way you are.* Oxford: Oxford University Press.

Nordlund, Marcus. 2007. *Shakespeare and the nature of love: Literature, culture, evolution.* Evanston, IL: Northwestern University Press.

Oatley, Keith. 2011. *Such stuff as dreams: The psychology of fiction.* Chichester, UK: Wiley-Blackwell.

Oatley, Keith, Raymond A. Mar, and M. Djikik. 2012. The psychology of fiction: Present and future. In I. Jaén and J. Simon, eds., *The cognition of literature.* New Haven, CT: Yale University Press.

Parsons, Keith M. 2003. *The science wars: Debating scientific knowledge and technology.* Amherst, NY: Prometheus Books.

Pinker, Steven. 1997. *How the mind works.* New York: Norton.

———. 2007. Toward a consilient study of literature. *Philosophy and Literature* 31: 162–178.

Plutchik, Robert. 2003. *Emotions and life: Perspectives from psychology, biology, and evolution.* Washington, DC: American Psychological Association.

Richerson, Peter, and Joe Henrich. 2012. Tribal social instincts and the cultural evolution of institutions to solve collective action problems. *Cliodynamics: The Journal of Theoretical and Mathematical History* 3. irows_cliodynamics_12453. Retrieved from: http://escholarship.org/uc/item/981121t8

Salmon, Catherine, and Don Symons. 2004. Slash fiction and human mating psychology. *Journal of Sex Research* 41: 94–100.

Saucier, Gerard, and Fritz Ostendorf. 1999. Hierarchical subcomponents of the Big Five personality factors: A cross-language replication. *Journal of Personality and Social Psychology* 76: 613–627.

Saunders, Judith P. 2007. Male reproductive strategies in Sherwood Anderson's "The Untold Lie." *Philosophy and Literature* 31: 311–322.

———. 2009. *Reading Edith Wharton through a Darwinian lens: Evolutionary biological issues in her fiction*. Jefferson, NC: McFarland.

———. 2012. Female mate-guarding in Lawrence's "Wintry peacock": An evolutionary perspective. 39: 69–83.

Scalise Sugiyama, Michelle. 2001a. Food, foragers, and folklore: The role of narrative in human subsistence. *Evolution and Human Behavior* 22: 221–240.

———. 2001b. New science, old myth: An evolutionary critique of the Oedipal paradigm. *Mosaic* 34: 121–136.

———. 2005. Reverse-engineering narrative: Evidence of special design. In Jonathan Gottschall and David Sloan Wilson, eds., *The literary animal: Evolution and the nature of narrative*, 177–196. Evanston, IL: Northwestern University Press.

Schore, Allan N. 2003. *Affect dysregulation & disorders of the self*. New York: Norton.

Smee, Sebastian. 2009. Natural-born thrillers. *Australian Literary Review*, May 6, 17.

Smith, Barbara Herrnstein. 2006. *Scandalous knowledge: Science, truth and the human*. Durham, NC: Duke University Press.

Snow, Charles P. 1993. *The two cultures*. Cambridge: Cambridge University Press.

Sokal, Alan D., and Jean Bricmont. 1998. *Fashionable nonsense: Postmodern intellectuals' abuse of science*. New York: Picador.

Spolsky, Ellen. 2009. The centrality of the exceptional in literary study. *Style* 42: 285–289.

Swirski, Peter. 2006. *Of literature and knowledge: Explorations in narrative thought experiments, evolution, and game theory*. Abingdon, UK: Routledge.

———. 2010. When biological evolution and social revolution clash: Skinner's behaviorist utopia. *Evolutionary Review: Art, Science, Culture* 1: 18–23.

Tooby, John, and Leda Cosmides. 2001. Does beauty build adapted minds? Toward an evolutionary theory of aesthetics, fiction and the arts. *Substance: A Review of Theory and Literary Criticism* 30: 6–27.

Vandermassen, Griet. 2005. *Who's afraid of Charles Darwin? Debating feminism and evolutionary theory*. Lanham, MD: Rowman & Littlefield.

Wade, Nicholas. 2006. *Before the dawn: Recovering the lost history of our ancestors*. New York: Penguin.

Weinberg, Steven. 2001. *Facing up: Science and its cultural adversaries*. Cambridge, MA: Harvard University Press.

Wilson, David Sloan, and Edward O. Wilson. 2007. Rethinking the theoretical foundations of sociobiology. *Quarterly Review of Biology* 82: 327–348.

Wilson, Edward O. 1998. *Consilience: The unity of knowledge*. New York: Knopf.

Winkelman, Michael A. 2009. Sighs and tears: Biological signals and John Donne's "whining poetry." *Philosophy and Literature* 33: 329–344.

13 EXPERIMENTS WITH EXPERIENCE

CONSILIENT MULTILEVEL EXPLANATIONS OF ART AND LITERATURE

Brian Boyd

Science and art are naturally distinct but also naturally consilient. If science is observation, hypothesis, experiment, so, too, is art. According to their specialty, artists observe sounds, sights, or situations; they hypothesize about possible recombinations of these elements; and their art always experiments with the potential experience of audiences.

For those who enjoy and study art in general or literature in particular, consilience, in E. O. Wilson's (1998) sense, presents a twofold challenge: how to construct a seamless explanatory system that ranges from the hard sciences to the softest arts—and not merely to put in scientific language what we knew anyway—and how to bring those further reaches of knowledge to bear concretely on the arts: on the ways we create, enjoy, understand, and evaluate works of art, and enrich ourselves during the process.

Joseph Carroll, the most active evolutionary literary critic, seeks to build a comprehensive model of human nature as explained by our best evidence, and to bring that to bear on the models of human nature in literary works (Carroll 2004, 2011). I welcome as full and accurate a model as we can get of human nature, as a goal, an asymptote, as I also welcome works of literature that offer rich models of what it is to be human. But a great deal of literature—literature that's not high art or large scale, not epics, say, but lyrics and comics—aims much less at constructing such models than simply at engaging minds. The search for the comprehensive is only a small part of art, and in some ways antithetical to the idiosyncratic experiments of art, just as most of the ferment in science focuses on particular problems—even if it has to coordinate with contiguous areas of knowledge—rather than on synthesizing all the known.

I therefore focus on three small contrasting experiments: two by Shakespeare, two consecutive sonnets, one famous, one not, the second written *in order* to contrast with the first (Shakespeare 2002 [1609]); and, contrasting in another way with these—with the work of the most classic of authors in one of the most classic of forms—a recent two-page comic by Art Spiegelman (Figure 13.1), the "Intro" to his *MetaMaus* (Spiegelman 2011). What do these authors experiment with and how do their experiments become data and delight for us? First, run these experiments on yourself (for Spiegelman's "Intro," see p. 226):

Sonnet 73

That time of year thou mayst in me behold
When yellow leaves, or none, or few, do hang
Upon those boughs which shake against the cold,
Bare ruined choirs, where late the sweet birds sang.
In me thou seest the twilight of such day
As after sunset fadeth in the west,
Which by and by black night doth take away,
Death's second self, that seals up all in rest.
In me thou seest the glowing of such fire,
That on the ashes of his youth doth lie,
As the death-bed whereon it must expire,
Consumed with that which it was nourished by.
This thou perceiv'st, which makes thy love more strong,
To love that well, which thou must leave ere long.
(Shakespeare 2002 [1609], 527)

Sonnet 74

But be contented when that fell arrest
Without all bail shall carry me away;
My life hath in this line some interest,
Which for memorial still with thee shall stay.
When thou reviewest this, thou dost review
The very part was consecrate to thee.
The earth can have but earth, which is his due;
My spirit is thine, the better part of me.
So then thou hast but lost the dregs of life,
The prey of worms, my body being dead,

The coward conquest of a wretch's knife,
Too base of thee to be rememberèd.
The worth of that, is that which it contains,
And that is this, and this with thee remains.
(Shakespeare 2002 [1609], 529)

These works are immediately accessible to their audiences—or if not immediately, for Sonnet 74, then on a second or third reading—without readers needing to know about evolution or cognition. So how can a consilient approach to literature, one that incorporates evolution (the why of human nature), and cognition (the how of human thought) help explain or enrich the experience of these works?

Evolution and cognition can explain literary works at multiple levels, but what parts evolutionary and cognitive findings play at most levels will, and should, vary considerably from case to case. For me, there should be no a priori methodology—just a willingness to bring a naturalistic understanding of human nature to bear on art and literature, and there should be no a priori conclusions. One of the most telling responses to the French-influenced literary theory of the late 20th century lay in the terms in which Frank Lentricchia, once the "Dirty Harry of literary theory," recanted. When he repudiated "Theory," he did so with a dismissal of its apriorism: "Tell me your theory, and I'll tell you in advance what you'll say about any work of literature, especially those you haven't read" (Lentricchia 1996, 60, 64). For consilient literary study, I would propose instead: "Give me your literary example and I'll go away and think how evolutionary and cognitive ideas can enrich our understanding and response."

Nevertheless, I propose we first import into literary analysis some generally applicable principles, obvious in evolutionary biology—*problem-solution models* and *cost–benefit models*— and then highlight two other terms richly relevant both to the evolution of human cognition and to art: *pattern* and *attention.*

Problems entered the world with life, beginning with the central problem of life, preserving the complexity that has already been achieved, in the face of all the forces threatening any organization of matter intricate enough to qualify as living—as able, that is, to sustain and reproduce its own complexity by finding energy and resisting physical wear and tear, parasitism, predation, competition within and beyond the species. The problem of preserving achieved complexity encompasses all the problems an organism faces in surviving and in enabling, through reproduction, the survival of its genetic constituents.

Biological structures solve problems, and better *solutions* to problems shape the way life evolves (Popper 1999). Animals, in turn, become active problem solvers. Social animals solve problems more efficiently than solitary ones by sharing

(a) (b)

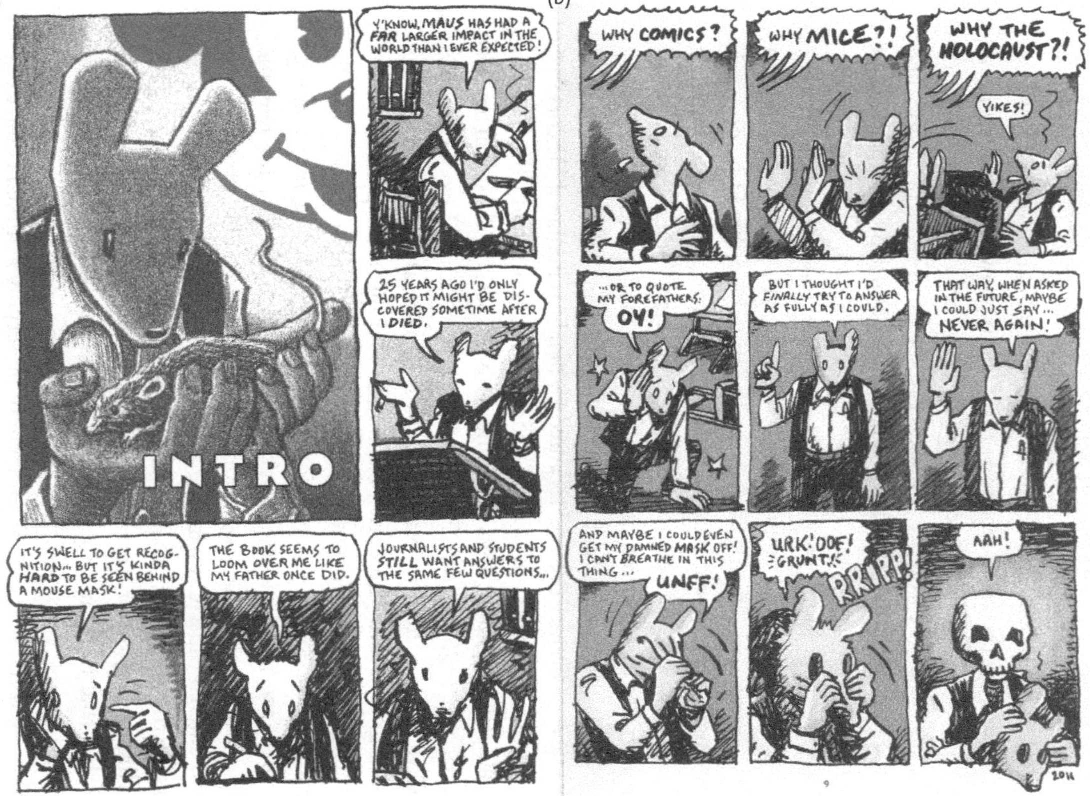

FIGURE 13.1 A panel from *MetaMaus*.

Source: Introduction to METAMAUS by Art Spiegelman. Copyright © 2011 by Art Spiegelman. Used by permission of The Wylie Agency LLC and Pantheon Books, a division of Penguin Random House LLC. Any third party use of this material, outside of this publication, is prohibited.

solutions (Wilson 2012), and extremely social animals can solve problems still further by specialization, by dividing problems up between different kinds of experts (Ridley 2010).

But every potential solution to problems within a given biological lineage will have *costs* as well as *benefits*. The trade-off between a solution's benefits and costs, relative to other available solutions, will determine whether it succeeds and whether, over many iterations, it shapes the way the species evolves. Costs and benefits arose with life, with the emergence of purpose in living organisms. Although they have long been recognized in toolmaking (and then engineering) and social exchange (and then economics), costs and benefits did not become fundamental to biological science until Darwin's theory of evolution made competition for limited resources the selective force central to explaining the diverse forms of life.

Problems and solutions, costs and benefits, pervade life—and art. Information allows organisms to respond appropriately to threats and opportunities in their environment, but analyzing information is costly. Slowly evolved modes of swift pattern recognition, such as recognizing outlines, shapes, movement, and direction, reduce the cost of interpreting information and allow animals to respond to it quickly, in real time. Brains therefore operate, observes neuroscientist Gerald Edelman, "not by logic but rather by pattern recognition"; "pattern recognition" is the "primary mode" of thought (Edelman 2006, 58, 103). We humans depend for our survival especially on our superior handling of information, and therefore on our capacity to recognize—and to find new ways of recognizing—*pattern*. That, I suggest, is the reason art emerged and became a human adaptation; art is cognitive play with pattern, developing our mental skills just as physical play evolved to be self-rewarding and compulsive because it develops key physical skills (Boyd 2009).

The world teems with potential information, but the neural space that can process such data is extremely costly and, therefore, despite the benefits it offers, very tightly limited. Just as pattern recognition offers a solution to processing abundant information in real time, *attention* offers a solution to the problem of limited neural space because it amplifies the processing of task-relevant information and inhibits the task-irrelevant. To cope with the world's plethora of potential information, automatic processes and conscious attention together sift for the information, and the patterns within it, most relevant for current purposes: for a species' customary range of purposes or an individual's purposes in a particular time and place.

How do these principles relate to art, including literature? Joseph Carroll worries that a problem–solution model when applied to art is only a metaphor (pers. comm. May 24, 2009). I suggest it is not; it is a causal chain, unbroken from the origins of life.

As a late part of this chain, humans have evolved to be motivated to engage in art, just as we and other animals engage in play, to improve performance in behaviors too flexible and context sensitive to be shaped by closed genetic programs. In the same way that play improves physical performance, so the cognitive play of art improves our processing of the kinds of information patterns central to us: sight, sound, and all forms of social information. Art, in other words, helps solve for us the problem of amplifying the advantages of our inclination to information.

Just as play works by being compulsive, by inviting repeat engagement and therefore incremental refinement, so does art. We normally have other demands on our time, resources, and attention, of course, and the option of rest and recovery when free of these demands. To benefit us art has to engage and hold our attention against competing demands, and to stir our responses, because emotion will amplify its enduring neural impact. Artists face the problem of maximizing what their art can earn within the marketplace for attention, and audiences face the problem of maximizing the value to them of what they direct their attention to, even in moments of relatively low urgency. Unlike vaguer models, a problem–solution model focuses on artists and audiences as active agents in a world of competing demands and opportunities. And just as we understand the social world in terms of the problems others seek to solve by their actions, we understand the world of a story in the same way, in terms of the goals and outcomes of characters' actions, and the role of the storyteller, in terms of the problems he or she faces and attempts to solve.

These general principles can therefore allow us to become very specific, to zoom in on the practices and products of art. We need to understand artists and audiences as problem solvers—artists as specialists trying, before all else, to solve the problem of engaging the attention of potential audiences through their works (and keeping invention costs down while they do so), and audiences as trying to maximize the benefit *for them* of the time and effort they expend while attending to particular works of art, when they could be pursuing practical goals or responding to other appeals to their disposable time.

If specialized artists can earn attention, their art offers them obvious benefits in terms of prestige and status. The benefits seem less obvious for audiences or those who engage in the arts more casually. Yet, if they were not real, a species-wide activity that usurps so much time and effort would have been bred out by those less inclined to engage in an activity with no objective benefits (Boyd 2009)—hence my explanation of art in terms of the benefits of cognitive play.

Artists, especially ambitious ones aiming beyond their own time and place, cannot fully know what audiences might want to attend to. Yet they can know enough of common human nature, of evolved preoccupations and expectations, to intuit what will have a good chance to catch and hold the attention even of more or less remote audiences. Much evolutionary literary criticism has focused precisely here, on the common human nature appealed to in art. But artists' natural appeal to common human nature also means they run the risk that their audiences will succumb to habituation, to the loss of attention to repeated or prolonged stimuli. Artists need to experiment to keep combating habituation and refreshing attention (Martindale 1990).

These key evolutionary and cognitive principles underlying art's experiments with experience permit very fine-grained analyses of particular works of art, when we adopt multilevel explanations.

Joseph Carroll has offered a three-level model of human experience, addressing the humanly universal, the culturally local, and the individual (Carroll 1995, 150). I would adapt this and expand it to at least six levels (Figure 13.2), including the *global* (principles beyond the merely human, within which human and other species operate), the *human* (operating similarly across cultures and individuals), the *local* (in culture, time, place, and circumstance), the *individual* author or audience member, the particular *work* (single artistic works as particular problem–solutions), and the *detail*. The degree to which any of these levels of potential explanation becomes actively relevant depends on the questions we want to answer.

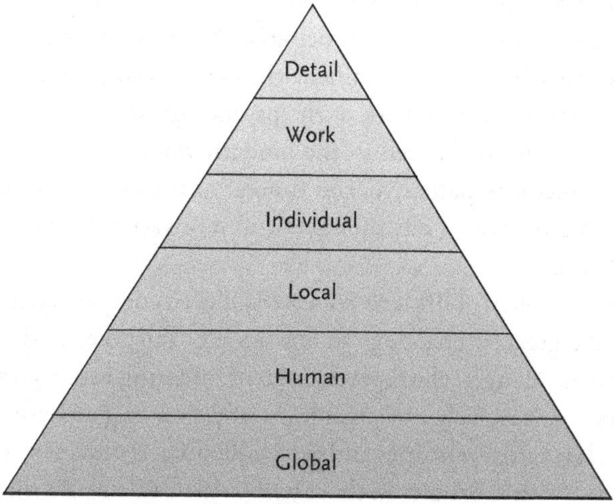

FIGURE 13.2 Levels of consilient explanation.

The *global* level includes principles such as *variation and selection*, and the capacity many species have for *sociality and social learning*: the general theoretical question, for instance, of memes as possible analogs of genes and as relevant to the differential survival of artistic works; the role of social learning in, for instance, food search or predator detection or, at a much more refined level, artistic traditions and conventions. But principles as general as variation and selection or social learning apply to ants as well as apes. We will find more causal precision as we move up the pyramid from base to apex.

The *human* level includes *our predisposition to art*. The question of whether art is adaptive has preoccupied evolutionary researchers in the arts but is only one question among many worth raising. I have proposed that art *is* adaptive, that art, as cognitive play with pattern, helps refine brains in key mental modes, in the case of fiction especially in terms of social cognition; other proposers offer other proposals, and none has yet reached the level of empirical testing (Dissanayake 1988, 1992; Pinker 1997; Dissanayake 2000; Miller 2000; Scalise 2001; Tooby and Cosmides 2001; Carroll 2004; Scalise 2005; Boyd 2009; Dutton 2009; Carroll 2011; Gottschall 2012). But in any case, most of us will assume the value and impact of art, without the need for scientific confirmation.

Another question at the human level, more sharply relevant to the particular subject matter of the two Shakespeare sonnets and Spiegelman's comic, is: What role does our consciousness of death play in human thought and feeling? Imagination and the ability to shift perspectives in time, place, and modality offer many advantages, first in devising alternative courses of action, then in testing them, then in allowing us to move further into new social, technological, scientific, and creative possibilities. But the evolution of an imagination as rich as ours also brought with it an unavoidable ability to imagine our own death and the deaths of those closest to us—our being bereaved, our being dead, our being no longer. We may think we focus sufficiently on life to keep anxiety about death in its place, but the empirical work of the psychological subfield of terror management theory shows that, even in the modern world, with greater longevity and specialized medical, palliative, and funerary care, even unrecognized hints of death impact strongly on our thoughts and motivation (Landau et al. 2007; Burke et al. 2010).

If the prime problem of life is to preserve achieved complexity, it is no wonder that death is the greatest challenge to our sense of purpose, or that apart from evading imminent danger, the reproduction of offspring is our greatest motivator. Art offers a particularly powerful form of preserving achieved complexity. Because it can last across centuries and even millennia, it provides a potent way to assuage anxiety about mortality in those with artistic talent and thereby to motivate the intense effort needed to secure artistic longevity (Landau et al. 2010).

For humans, sharing attention has evolved to matter uniquely (Tomasello 1995, 2008; Tomasello et al. 2005). In reading a lasting work of literature we can share attention with the writer and other readers—past, present, and future—in a way that allows us some sense that we belong intimately together to a human world not completely subject to time, death, and decay (Boyd 2012). That certainly becomes a recurrent subject in Shakespeare's Sonnets; and Spiegelman both explains his motivation for the 13 years of effort that *Maus* took ("I'd only hoped it might be discovered after I *died*" [Spiegelman 2011, 8]) and toys with the fear of death, conquers it, as were, by facing it and playing with it. Art, as here, can create work worth human attention for a long time—human instances of mastery that last, that have been designed to appeal to other humans, including those still to come, and that therefore embody a trust that humanly achieved complexity can survive even in the face of death.

Of course death and our anxieties about death have hardly been unknown to people in general or artists in particular. But the empirical work of terror management theory adds objective detail about our anxieties, even when we might seek to downplay it, whereas Michael Tomasello's stress on the unique role of human shared attention helps explain the emotional charge of art that allows us to engage with the minds of artists long since dead.

After the global and the human levels of explanation, we come to the *local* level: local in terms of culture, history, geography, politics, economics, and so on. This level has been the overwhelming focus of literary research for the past 35 years, with the local cultural contexts often seen as playing a more decisive causal role in literature than even individual authorship.

Evolutionary literary criticism has often been reproached for being unable or unwilling to factor in local conditions. This is a knee-jerk reaction, based on our interest in human universals rather than on the evidence of our work. Evolutionary literary critics naturally assume—it is, after all, an elementary principle of evolutionary theory—that different behaviors will be elicited in different cultural and social as well as different natural ecologies. Explanations that focus only on the local, on the other hand, can overlook or exclude the "universal" features with which local conditions interact.

The local factors most immediately relevant to works of art are the local artistic manifestations of the power of social learning: artistic *modes* like verse, *genres* like lyrics, *forms* like sonnets, *conventions* like speech balloons, and *traditions* of reading and exegesis. Shakespeare wrote sonnets, a verse form invented during the 12th century and still enduring, I have argued, because its structure allows for complex patterns; it sets up and shifts precise expectations that can then be further varied or violated (Boyd 2012, 32–34). Sonnet structure reflects a general cognitive principle; animals seek out pattern, because it allows information to be processed rapidly, in real time, and humans crave pattern in open-ended ways. Sonnets offer, within a compact scope, a saturation of partly conflicting and therefore unpredictable and habituation-resistant patterns (Boyd 2012, 9–34).

Shakespeare was writing not just single sonnets, but within the larger form of the sonnet sequence, a form invented by Petrarch during the 14th century and remaining popular until the early 17th century. The sonnet sequence, I argue, is in part a solution to the problem of artistic scale (Boyd 2012, 65). Scale matters in art; imperfect epics easily outweigh perfect epigrams. The most ambitious lyrics poets could earn a new kind of attention and prestige through linking their sonnets into the larger structure of sonnet sequences.

In the form that Petrarch established and that ensured his undying fame, a sonnet sequence features a male poet wooing a single female whose rejection of the poet's advances does nothing to diminish the continued expression of his love. We can explain the appeal of the sonnet sequence in part through an evolutionary account of the asymmetry between sperm and egg, the massive production of sperm and the sparing production of eggs, and the differential parental investment of male and female (Trivers 1972, 2002). Females have reasons to be choosy, because they have to invest so much more in their offspring than males. Females, therefore, tend to resist male advances, and the more they resist, the more males need to persist (Seabright 2012). That general evolutionary logic, coupled with the particularly high Christian evaluation of female chastity, explains the appeal of the Petrarchan sonnet sequence, whereas internal structural problems of the Petrarchan sequence—especially the fact that it leads to a sexual stalemate—explain its eventual exhaustion (Boyd 2012, 64–65).

Spiegelman works in comics, a form invented at the end of the 19th century. A very self-aware artist, he calls comics "co-mix" (Spiegelman 1999, 74)—a mix of the verbal and the visual—and therefore, as he notes, appealing to our richest information modes, sight, and language. In comics, as he also stresses, the visual can be reduced to the essential, the near-iconic. Comics' comprehension costs, therefore, tend to be low, but composition costs can be high; storytellers can tell stories much more efficiently in just words than in the combination of words and static pictures. But amid the New York newspaper wars waged by Joseph Pulitzer and William Randolph Hearst at the end of the 19th century, Richard Outcault invented a drawn character whom he combined with words within the picture. The Yellow Kid became such a hit—the high benefit/low-cost ratio immediately appealed to audiences—that he boosted the circulation of Pulitzer's newspapers sufficiently for Hearst to poach Outcault and for comics to be born (Boyd 2010).

The success of comics as a medium can be explained in terms of general cognitive features: our predisposition for narrative and fiction; the dedicated neural architecture for processing outlines, which served cave painters and now helps us track comics characters effortlessly; the pop-out effect of color in human visual processing; our predisposition for humor, to which most early newspaper comics appealed—hence the very name. These pan-human features interacted with local features such as mass education, huge print runs, and the intense time competition for attention in newspapers that made comics such a powerful selling point. Comics can offer relatively high benefits for very low comprehension costs: they fuse stripped-down, already prefocused visual information, and highly focused verbal information, almost exactly as in real-life social situations, but without their information redundancy. Evolved cognitive features interacting with local circumstances led to the invention of comics, and their swift success in their original context and many others can be explained by their regular ratio of low comprehension cost to high subjective benefit.

To explain *Spiegelman's* kind of comics we need still more local circumstances. For instance, the invention of web printing made it possible to print affordably in much smaller print runs, allowing an experimental comics tradition to emerge during the late 1960s. With printing costs lower, comics artists could afford to appeal not to mainstream audiences, but to niche audiences.

Now we can tighten our focus further, to the level of the *individual*. No one creates works of art without being shaped by their local context, including the context of artistic modes and instances available to them. But roughly similar contexts exist for everyone in the same time and place, and only a vanishingly small proportion of individuals there create art that earns lasting attention.

We naturally expect individual differences between person and person; science can help explain them as a consequence of genetic variation and epigenetic development, and of the development of specialized expertise within accommodating social structures that allow for the high degrees of specialization that make it possible to intensify interpersonal differences in skills (Simonton 1999; Ridley 2010). We naturally expect individuals to act differently according to their different situations and purposes. We can formalize this through the idea of *problem situations* (Popper 1992 [1974]), and the *costs and benefits* of different choices to clarify individual behavior still further.

I argue that, for artists, the core problem is usually to earn the attention of audiences—perhaps especially that of noble or wealthy patrons, or a high social class, or the widest possible mass audience, or a prestigious audience of cognoscenti, or a certain age or gender cohort (Boyd 2009). The role of attention has been recognized by artists all along, and by critics since at least Aristotle, and for instance by Hollywood screenwriting teachers, but it has tended to be overlooked by academics disposed, trained, and even paid to attend to the artworks in their specialty. Academic criticism has tended to focus on meaning and take attention for granted, as successful artists never can.

Shakespeare began to write his sonnets, recent evidence suggests, in 1594 to 1595 (Jackson 1999, 2000, 2001, 2002a, 2002b). By that time he had already established his preeminence in the foremost narrative modes of his day: in drama, in tragedy, comedy, and history; and in narrative poetry, both comic and tragic. I hypothesize that at this point this ambitious poet sought to attain a matching preeminence in the most competitive lyric mode of his day, the sonnet sequence, and, moreover, as someone already triumphant in narrative forms, that he sought to do so especially by showing how much he could do with pure lyric, without the narrative elements that other sonnet sequence poets occasionally allowed (Boyd 2012, 67–68). He could show to a new degree how he could earn attention without narrative—and he did; his sonnets are the most successful lyrics in Western tradition.

Spiegelman, too, faced a unique problem situation caused by his own success. He had become the most radical of experimental comics artists, and the more he experimented, the narrower his audience became. He was stretching the bounds of comics possibilities, but losing an audience in doing so—not what he wanted for himself or the medium he saw so much scope for. He also had a theme close to his heart, his parents' experience before, in, and after Auschwitz, and realized that this would offer an opportunity to show what comics could do with a large, intense, and decidedly noncomic subject. He could also use the unique medium of comics for a general critique of the racism that led to the Holocaust. In his masterpiece, the two-volume *Maus*, which took him 13 years, he depicts Jews as mice, Germans as cats, Poles as pigs (Spiegelman 1986, 1991). He takes the racist metaphor that other human "races" are tantamount to other species, and explodes it in every frame, because we cannot help seeing his characters as completely human in gesture, posture, actions, emotions, and circumstances, despite their animal heads, and the mice as overwhelmingly sympathetic, despite their honestly rendered foibles and their usual status to humans as vermin.

Spiegelman's solution was so successful that *Maus* quickly became the most critically acclaimed comic of all time, and a bestseller in many languages. But that in turn created a new problem: for the past 20 years, Spiegelman has lived under the shadow of *Maus*'s success. (To make a general point: there can often be more purchase on authorial aims by focusing on these individual problems, at specific career moments, in specific situations, than on supposing authors always focus on cosmic human problems.) To satisfy once and for all the intense curiosity *Maus* elicited, Spiegelman decided to turn the in-depth interviews of a gifted young scholar, Hillary Chute, into the basis of a book, *MetaMaus*, mostly words and photographs, transcripts and documents, but also his own comics, new and old, relating to *Maus*, to his development as an artist, and to the medium in which he worked (Spiegelman 2011). The "Intro" solves the problem of introducing *Maus* and its impact, and the volume *MetaMaus*, and the problems of showcasing Spiegelman's own voice and his handling of comics and the mouse metaphor, and of demonstrating the art, wit, and resonance possible even in a mere introduction.

A problem for artists as for everyone, even every organism, is to reduce the costs of their efforts while raising the benefits. Like all social species, we reduce search costs through social learning, as Shakespeare did by choosing the established forms of the sonnet and the sonnet sequence. But imitating others in art runs the risk of audience habituation, the waning of audience attention through repeated or sustained stimuli (Martindale 1990). Surprising expectations will earn fresh attention. Rather than extolling a fair lady who resists his poetic entreaties, Shakespeare began his sonnet sequence by introducing a Mistress (the so-called Dark Lady) who not only is dark in feature rather than the expected fair, but also, and much more important, does not resist him and even solicits other men.[1]

That surprised in Shakespeare's own day, and still does (he calls her "a bay where all men ride," Sonnet 137.6). But Shakespeare's bold solution to the problem of earning attention also entailed its own problems. Although the conventional sonnet sequence tends to find itself in a stalemate, as the persistently adoring poet repeatedly exhorts the resistant lady, Shakespeare finds little poetic reason even to persist in addressing a woman who does not resist. And the mixture of desire and disgust appropriate to such a relationship does not yield the emotional appeal of a more conventional sonnet sequence situation. The sonnet sequence addressed to the Mistress fizzled out quickly, and Shakespeare did not publish it for another 14 years—a long time in his 24-year writing career—until he had found and developed a better solution to surprising expectations.

We have begun to tighten the focus still further to the particular level, the level of the single *work*. As he worked on the sonnets, Shakespeare found another, more radical and more fertile, solution to the problem of habituation to the sonnet sequence formula: instead of addressing a dark woman who does not resist, and who therefore does not prompt his persistence, he addresses a decidedly fair *man* who resists all the women who would be ready to have him. As poet, Shakespeare persists in trying to persuade the Young Man to overcome the risk of disappearing without a trace—of losing to time and death all his current perfection, all his achieved complexity, to reintroduce those terms—if he does not reproduce. This provides the basis for the published sequence's first 17 sonnets, which segue into another 109 sonnets that continue to idolize the Young Man, as the sonnets to the Mistress tend to demonize her. In many of the sonnets that follow, Shakespeare acknowledges that if he cannot help the Young Man to reproduce, he can at least offer him, through the force of his art, another way to withstand time and death.

This radical solution proves much better, but it is also off-putting to some. Shakespeare addresses the Young Man in terms of love, which some readers in the homophobic past have found disturbing. Yet most of the first 126 sonnets do not specify the sex of the addressee (although if they *do* specify, it's always male, just as many of the last 28 sonnets do not specify the sex of the addressee, but if they do specify, it's always female). So there's a complicated flip possible, a singular doubleness: each of these sonnets can be taken on its own, and if the sex is unspecified readers can make the default assumption that a male addressing another person in terms of love is addressing a female, but if the sonnet is read within its sequence we may wish to see it as male addressing male (Boyd 2012, 101–110). When we read Sonnets 73 and 74 in context, they belong to the sequence of the Young Man; read on their own, they could refer to a beloved of either sex.

Love almost guarantees attention, from sonnet sequences to pop songs. But the very popularity of the subject risks habituation, not least in Petrarchan sonnet sequences, where love forms the almost invariable premise. In Sonnet 73—and we have now reached the level of the *detail*—Shakespeare experiments with expectations by not even introducing love until the last moment, the final couplet, and by the sonnet's unusual take on the subject: he praises the other person's love for him because it can withstand even the certain prospect of his own looming death. Love at its best can persist even in the face of time, decay, and irretrievable loss.

The sonnet focuses on analogies, on pattern superimposed on pattern, the pattern of a human life superimposed on other patterns in time and life: the patterns of the seasons, of day and night, and of the dying—or what we might now call the heat exhaustion, the entropic loss—of fire. Shakespeare's Sonnet 73 appeals—it is one of his most loved and admired—because it takes these analogies, which seem a natural way for us to make sense of the life cycle, but introduces them in such novel and vivid forms: "That time of year thou mayst in me behold." Shakespeare also makes the most of sonnet structure, tying each image to one quatrain with its own rhyme patterns, and its own "In me thou seest" or equivalent. He offers us pattern on pattern on pattern, demarcated vividly and lucidly to reduce comprehension costs.

When we read the sonnets closely, an evolutionary and cognitive focus does not necessarily lead to a reading drastically different from those possible for responsive readers for the past few centuries. Indeed, it would be strange to claim that it did, that Shakespeare had written in a way that only knowledge that he could not possibly foresee would make his work fully comprehensible. But an evolutionary and cognitive focus can make things a little deeper, wider, and sharper, and can *explain* the effects the sonnets have on us, including why we value these effects.

Other evolutionary critics might well focus on other factors, but I would emphasize patterns, and the mind's predisposition to patterns, and the difference between the automatic convergence of patterns on the events of a story and the more uncertain correlation of patterns within the openness of a lyric (Boyd 2014). A sonnet can have tight rhyme patterns that tend to suggest preestablished structural patterns, but these can be played unpredictably with or against—by idea, emotion, and image—in an elegant dance of expectation and surprise. If we consider the mind's predisposition to patterns, we can answer a question that otherwise might remain unanswerable: if we don't take the practice of reading sonnets for granted, why *should* they appeal to us?[2]

I would emphasize, too, the openness of a lyric situation, and the sense of intimacy a reader can have with a lyric poet, possible because a lyric is not constrained by story. Shakespeare addresses the poem to a single *thou*, which we can see as an archetypal beloved, *or* the Young Man of the rest of this subsequence, *or* ourselves filling the placeholder of the pronoun. We have a sense of our eavesdropping on the poet at a moment of intimate address, or even stepping into the position of the beloved. Tomasello's work on the unique human motivation for shared attention (Tomasello 1995, Tomasello et al. 2005; Tomasello 2008), the mirror–neuron research of Rizzolatti and others (Iacoboni 2008), and the work of Goleman (2006) and others on emotional attunement help explain why this experience should matter to us. Terror management theory also helps explain why it should matter particularly across the great gap in time here, why we should value communion with an author long dead, and with other readers, past, present, and future. We can sense that, despite and across the obliterating effects of time, we can all direct and share attention in very much the same way, we can all engage in this experiment in experience and hold on to and return to this achieved complexity that can withstand time and individual extinction.

I would emphasize, too, the resonance and relevance of the theme of impending decay and loss. Usually language operates with speaker and hearers in the same time and place, where whatever is relevant to the situation will largely be shared by all (Sperber and Wilson 1996). Narrative, in contrast, supplies its own internal relevance, in terms not of a shared here and now, but of the unfolding situation, within the there and then, that hooks audience attention. Lyrics reflect neither the actual here and now of author or auditors nor the there and then of narrative characters. Lyric poets are completely unconstrained by situation, and therefore have to appeal to auditors' sense of relevance, without knowing in advance what that might be, and therefore often by focusing on something common and central to all human life. Auditors, for their part, if they want lyrics to matter for them, will allow implications to resonate to the utmost, to maximize the relevance *for them*. All the more so when lyrics move from oral delivery to print, so that poets may address unseen and unknown audiences in perhaps remote times and places.

I therefore read Sonnet 73 not radically differently from other good readers, but I can explain *why* what happens in reading a lyric *matters* in terms of the neural logic of attention and habituation; in terms of the cognitive and emotional costs and benefits of pattern perception; in terms of emotional attunement, and the impact of that on social cohesion and human existential anxiety; in terms of the relevance theory of Sperber and Wilson. I can also use the biological idea of organisms coping with their problem situations to clarify Shakespeare's situation as a writer at the point in his career where he began to compose a sonnet sequence. I focus not just on his milieu in general, or even on the local cultural phenomenon of the sonnet sequence, but also on what it meant for *him* to undertake a sonnet sequence at this point of his career. And we should note that his problem situation in Sonnet 73 was not that he was really entering the late autumn of his life: he was probably not yet 35 when he wrote it.

Sonnet 74 offers quite a different experience, and has had quite a different fate. It has never been one of Shakespeare's famous sonnets, and we can propose a number of reasons why. Like Sonnet 73, it, too, focuses on love and death, and a third theme: art. But it lacks the vivid and accessible imagery of Sonnet 73; it presents something of a riddle to readers, rather than the open invitation of 73; it contrasts the perishable body with the "*this*," the imperishable poetic line, in a way that we may well find unconvincing and not particularly relevant to our own predicament. Even if we accept that we will die, *we* are not in a position to leave immortal lines of verse behind us, and we may be reluctant to concede that everything associated with our body was but "the dregs of life . . . too base to be rememberèd."

But if we see the sonnet in terms of the problem situation that Shakespeare poses himself, it takes on a fascination of its own. Sonnet 74 is pointedly a riposte to Sonnet 73. Sonnet 73 ends "To love that well, which thou must leave ere long." Sonnet 74 starts "But be contented when that fell arrest/Without all bail shall carry me away." When death forces me to move on and to leave you, there will still be something of me, and that most important part, that will stay with you—namely, this poem expressing the best that my mind can do and my heart can feel. This alone can transcend death.

As so often happens, Shakespeare solves creative problems by flipping an idea over to its opposite to spur invention and invite attention: in the sequence as a whole, from the standard fair lady who adamantly resists the poet, to a *Dark* Lady who *resists no man*, then to a *fair* Young *Man* who at first *resists all women*; in this sub-subsequence of two, from a sonnet that ends with the beloved left alone, without the poet, to a sonnet with the beloved left with what matters of the poet and outlasts death, this poem of love; from a sonnet vivid with visual imagery to a sonnet that keeps to abstractions; from a sonnet that ends "This . . . that . . . thou" to a virtuoso reply, with three *thats* and two *thises* in the couplet, in lines that would mean nothing elsewhere, but clinch the poetic case here.

(A slight digression. For those who focus exclusively on meaning, Sonnet 74's chief interest may be what it suggests about Shakespeare's thinking at this point. The poet uses the Christian contempt of the corruptible body to contrast with the endurance of art in a way that resolutely avoids anything of *Christian* consolation—and offers, indeed, rather meager consolation to any of us. Nevertheless, Shakespeare's expressing the contrast between the evanescence and worthlessness of the mortal body and the permanence and worth of this poem, as a product of his spirit rather than his flesh, seems almost less his object than forming a sonnet to pair and contrast with Sonnet 73.)

The differential success of different works of art seems particularly pertinent to a Darwinian perspective on art. The experiments of life do not always work, and may have very different ranges of effects; the same holds true for the experiments of science and art. There are good reasons why Sonnet 73 appears in many anthologies and Sonnet 74 in almost none. The weaknesses of 74 can be formulated in some of the terms I have brought to bear on 73: that it addresses a problem too localized, a contrast to the poet's *leaving* thee in death in 73 with this new sonnet's *staying with* thee past death; that it fails to find the right relevance and resonance to most people's experience, even if it does deal with the universal and troubling recognition of death; that it doesn't invite emotional attunement. But at least looking at the poem in terms of problems and patterns helps us understand exactly what Shakespeare was doing.

In our third "experiment with experience," Spiegelman has a common enough problem situation, writing the introduction to the book that follows, but he turns it into a much richer problem than usual, and therefore into a miniature work of art in its own right. He addresses us directly, as an introduction does, but addresses us as the cartoon character that he has made himself in *Maus* and that is part of the real-life problem that he wants to introduce: that he is and will always be known as the maker of *Maus*. Here again we slide from the level of the particular *work* to the *detail*. He introduces himself as cartoon character, as autobiographer, as comics artist, as mouse (both as Jew, and as a character in his own treatment of the fate of Polish Jews), and as explorer of the possibilities of comics. The first two frames (Figure 13.1) show four different comic possibilities of "mousification," the literal mouse in his hands, the Mickey Mouse on the wall—the kind of commercial comics art he does not want people to think sums up the medium's potential—the stylized but detailed Mouse he is himself in this title frame, and the sketchy Mouse he remains in the next panel and the rest of the comic. He will discuss and demonstrate these different possibilities elsewhere in *MetaMaus*—after this opening frame, so decidedly meta-mousy in a single image.

He holds our attention, and pays us the compliment of expecting our comprehension, by the interplay between himself as metaphoric mouse and real human individual that underpins the fascination of *Maus* itself. He takes advantage of our predisposition to understand experience visually and concretely, and to understand others especially in face-to-face encounters, in representing himself looking up from his drawing board and talking to us as people interested in *Maus*. He shows the resources of his medium everywhere: in the triple-pronged jagged-edged speech balloons, stylizing endless, jarring repetition; in the slapstick of the successively more intrusive questions eventually knocking him right off his drawing-board chair; in the comic discrepancy between the fulsome preamble, "or to quote my forefathers" and the comically curt "Oy"; the black-comic echo of the hope "Never again" so often expressed about the Holocaust in his own "never again" to these questions to him about his report on the Holocaust.

His predicament of having to live with and be defined by a much earlier success may not be that of most, but he manages to end his account of this individual fate with something that resonates with our common fate. He reports his initial hope, at a time when his work was being read by fewer and fewer, that *Maus* might be discovered some time after he died—in keeping with the motivation that terror management theory documents, of death sharpening our desire to leave something socially valued behind us. Ironically, he faces an opposite situation now, *Maus* staying with him until he dies. The theme of death, of course, also hovers over the subject matter of *Maus* and the questions he reports about the Holocaust. The last tier brings it all home with apt ironic finality: even *MetaMaus*, despite its aim to stave off further questions, won't allow him to tear off the Maus mask, until he dies, until he stands exposed as what we all become, the skull beneath the skin. His introduction has focused on his own peculiar circumstances, but it suddenly ends in the last frame with a universal human destiny. At the same time, it's also an ironic undercutting of the appeal of artistic immortality, an undercutting in its in-your-face or in-your-skull humor that allows us another way of coping with death. And like all his work, it's a demonstration of the power of comics, through an effect possible only in comics.

The closer we get to literary particulars, to the unpredictables of a work or its details, the less predictable the relevant evolutionary and cognitive plug-ins, *except* for common heuristics like problems and solutions, costs and benefits, attention and pattern. But there can be plug-ins at any level, like the asymmetry between sperm and egg ultimately explaining the structure and success of the Petrarchan sonnet sequence, or our evolution for face-to-face encounters, or our alertness to outlines and to animal features.[3]

Does an evolutionary and cognitive perspective trump the response of other attentive readers? Does evolutionary and cognitive language only put the intuitively obvious into an elaborate framework that delays us from reaching the literary interest we want? Or does it, as I suggest, offer a deeper, and a more coherent, consilient explanatory framework, with no loss of particularity, precision, and subtlety?

Notes

1. For the evidence that Shakespeare began his Sonnets by composing the Mistress poems, see Jackson [2002a, 2005].
2. The relationship between pattern, expectation, and reward in literature needs further investigation at the neural level. As the work of Read Montague, Peter Dayan, and Wolfram Schulz has shown, dopamine neurons of the ventral tegmental area and substantia nigra predict rewards (Schulz et al. 1997). "What's interesting about this system is that it's all about *expectation*. Dopamine neurons constantly generate patterns based on experience; if this, then that. The cacophony of reality is distilled into models of correlation. And if these predictions ever prove incorrect, then the neurons immediately readjust their expectations" (Lehrer 2008). David Huron (2006) has developed an influential cognitive and evolutionary theory of musical response based on expectation that he generalizes to all expectation and that should prove relevant to the patterns, expectations, and pleasures of both narrative and verse.
3. Joseph Carroll responded to an oral presentation of this paper by describing it as description rather than explanation. By my count I offer 75 explanatory proposals for the works under consideration, at all the six levels I suggest, although because I focus on three brief test texts, most of the explanations come at the more pointed end of the pyramid, at the local, individual, particular, and detail levels.

References

Boyd, Brian. 2009. *On the origin of stories: Evolution, cognition, and fiction*. Cambridge, MA: Belknap Press of Harvard University Press.

———. 2010. On the origins of comics: New York double-take. *Evolutionary Review: Art, Science, Culture* 1: 97–111.

———. 2012. *Why lyrics last: Evolution, cognition, and Shakespeare's sonnets.* Cambridge, MA: Harvard University Press.

———. 2014. Story versus verse: Open versus convergent pattern. *Sewanee Review* 121: 34–54.

Burke, Brian L., Andy Martens, and Erik H. Faucher. 2010. Two decades of terror management theory: A meta-analysis of mortality salience research. *Personality and Social Psychology Review* 14: 155–195.

Carroll, Joseph. 1995. *Evolution and literary theory.* Columbia: University of Missouri Press.

———. 2004. *Literary Darwinism: Evolution, human nature, and literature.* New York: Routledge.

———. 2011. *Reading human nature: Literary Darwinism in theory and practice.* Albany: State University of New York Press.

Dissanayake, Ellen. 1988. *What is art for?* Seattle: University of Washington Press.

———. 1992. *Homo aestheticus: Where art comes from and why.* Seattle: University of Washington Press.

———. 2000. *Art and intimacy: How the arts began.* Seattle: University of Washington Press.

Dutton, Denis. 2009. *The art instinct: Beauty, pleasure, and human evolution.* New York: Bloomsbury.

Edelman, Gerald. 2006: *Second nature: Brain science and human knowledge.* New Haven, CT: Yale University Press.

Goleman, Daniel. 2006. *Social intelligence: The new science of human relationships.* New York: Bantam.

Gottschall, Jonathan. 2012. *The storytelling animal: How stories make us human.* New York: Harcourt Houghton Mifflin.

Huron, David. 2006. *Sweet anticipation: Music and the psychology of expectation.* Cambridge, MA: MIT Press.

Iacoboni, Marco. 2008. *Mirroring people: The new science of how we connect with others.* New York: Farrar, Straus and Giroux.

Jackson, MacDonald P. 1999. Rhymes in Shakespeare's *Sonnets*: Evidence of date of composition. *Notes and Queries* June: 213–219.

———. 2000. Aspects of organisation in *Shakespeare's Sonnets* (1609). *Parergon* 17: 109–134.

———. 2001. Vocabulary and chronology: The case of Shakespeare's sonnets. *Review of English Studies* n.s. 52: 59–75.

———. 2002a. Dating Shakespeare's sonnets: Some old evidence revisited. *Notes and Queries* June: 237–241.

———. 2002b. The distribution of pronouns in *Shakespeare's Sonnets. AUMLA* 97: 22–38.

———. 2005. Francis Meres and the cultural contexts of Shakespeare's rival poet sonnets. *Review of English Studies* n.s. 56: 224–246.

Landau, Mark J., Sheldon Solomon, Tom Pyszczynski, and Jeff Greenberg. 2007. On the compatibility of terror management theory and perspectives on human evolution. *Evolutionary Psychology* 5: 476–519.

Landau, Mark J., Daniel Sullivan, and Sheldon Solomon. 2010. On graves and graven images: A terror management analysis of the psychological functions of art. *European Review of Social Psychology* 21: 114–154.

Lehrer, Jonah. 2008. A new state of mind. *Seed* August 8. http://seedmagazine.com/content/print/a_new_state_of_mind/ [Online].

Lentricchia, Frank. 1996. Last will and testament of an ex-literary critic. *Lingua Franca* September–October: 59–67.

Martindale, Colin. 1990. *The clockwork muse: The predictability of artistic change.* New York: Basic Books.

Miller, Geoffrey. 2000. *The mating mind: How sexual choice shaped the evolution of human nature.* New York: Doubleday.

Pinker, Steven. 1997. *How the mind works.* New York: Norton.

Popper, Karl. 1992 [1974]. *Unended quest: An intellectual autobiography.* London, Routledge.

———. 1999. *All life is problem-solving.* London: Routledge.

Ridley, Matt. 2010. *The rational optimist: How prosperity evolves.* London: Fourth Estate.

Scalise Sugiyama, Michelle. 2001. Narrative theory and function: Why evolution matters. *Philosophy and Literature* 25: 233–250.

———. 2005. Reverse engineering narrative: Evidence of special design. In Jonathan Gottschall and David Sloan Wilson, eds., *The literary animal: Evolution and the nature of narrative,* 177–196. Evanston, IL: Northwestern University Press.

Schulz, Wolfram, Peter Dayan, and P. Read Montague. 1997. A neural substrate of prediction and reward. *Science* 275: 1593–1599.

Seabright, Paul. 2012. *The war of the sexes: How conflict and cooperation have shaped men and women from prehistory to the present.* Princeton, NJ: Princeton University Press.

Shakespeare, William. 2002 [1609]. *Complete sonnets and poems.* Colin Burrow, ed. Oxford: Oxford University Press.

Simonton, Dean Keith. 1999. *Origins of genius: Darwinian perspectives on creativity.* New York: Oxford University Press.

Sperber, Dan, and Deirdre Wilson. 1996. *Relevance: Communication and cognition,* 2nd ed. Oxford: Wiley-Blackwell.

Spiegelman, Art. 1986. *Maus: A survivor's tale, I: My father bleeds history.* New York: Pantheon.

———. 1991. *Maus: A survivor's tale, II: And here my troubles began.* New York: Pantheon.

———. 1999. *Comix, essays, graphics and scraps: From Maus to now to Maus to now.* New York: Raw.

———. 2011. *MetaMaus: A look inside a modern classic.* New York: Pantheon.

Tomasello, Michael. 1995. Joint attention as social cognition. In C. Moore and P. Dunham, eds., *Joint attention: Its origins and role in development*, 103–130. Hillsdale, NJ: Lawrence Erlbaum

———. 2008. *The origins of human communication*. Cambridge, MA: MIT Press.

Tomasello, Michael, Melinda Carpenter, Josep Call, Tanya Behne, and Henrike Moll. 2005. Understanding and sharing intentions: The origins of cultural cognition. *Behavioral and Brain Sciences* 28: 675–735.

Tooby, John, and Leda Cosmides. 2001. Does beauty build adapted minds? Towards an evolutionary theory of aesthetics, function, and the arts. *SubStance* 20: 6–27.

Trivers, Robert. 1972. Parental investment and sexual selection. In B. Campbell, ed., *Sexual selection and the descent of man: 1871–1971*, 163–179. Chicago: Aldine.

———. 2002. *Natural selection and social theory: Selected papers of Robert Trivers*. New York: Oxford University Press.

Wilson, Edward O. 1998. *Consilience: The unity of knowledge*. New York: Knopf.

———. 2012. *The social conquest of earth*. New York: Liveright.

VI A CHALLENGE

14 THE LIMITS OF CONSILIENCE AND THE PROBLEM OF SCIENTISM

Massimo Pigliucci

The Two Cultures and the Science Wars

Knowledge, according to Plato, is justified true belief (Chappell 2009). If you believe, say, that Earth rotates around the sun, this belief counts as knowledge if two conditions hold: it is true, meaning Earth really does circle the sun and not, say, the other way around; and you are able to justify why you hold it, as opposed to simply repeating what other people told you. By this definition, we all know much less than we think we know. For instance, many people will tell you that matter is made of quarks, but assuming that is true (is it not strings?), only a handful of physics students can even begin to explain why we think that things are made of quarks. It is indeed a humbling experience to run quickly through a partial list of the things you believe and separate them into justified true belief on one side and mere opinion or hearsay on the other.

If we move from personal knowledge to the store of knowledge of humankind, we enter the realm of hot disputes between academic disciplines and, in particular, an ongoing war of words between what C.P. Snow (2012 [1959]) famously called "the two cultures": the humanities on one side, the sciences (natural and physical) on the other, with social science and philosophy caught somewhere in the middle. This is more than a turf dispute among a small cadre of academics (about which see, for instance, Labinger and Collins [2001]); it strikes at the core of what we mean by human knowledge and how we go about augmenting it and communicating it.

Snow (2012) arguably brought this debate into the open for the first time with his essay, originally published in 1959. He started out his career as a scientist and then moved to the humanities, famously recounting his dismay at the attitude of his newly acquired colleagues:

> A good many times I have been present at gatherings of people who, by the standards of the traditional culture, are thought highly educated and who have with considerable gusto been expressing their incredulity at the illiteracy of scientists. Once or twice I have been provoked and have asked the company how many of them could describe the Second Law of Thermodynamics. The response was cold: it was also negative. Yet I was asking something which is the scientific equivalent of: Have you read a work of Shakespeare's? (Snow 1959, 15–16)

That was more than half a century ago and, if anything, the situation got worse in some quarters of the humanities. Throughout the 1990s, for instance, postmodernist, deconstructionist, and radical feminist authors (think Michel Foucault, Jacques Derrida, Bruno Latour, Sandra Harding and company) wrote all sorts of questionable things about science, sometimes demonstrably without understanding what scientists actually do. Just for a taste, consider this statement by feminist philosopher Sandra Harding (1986): "I doubt that in our wildest dreams we ever imagined we would have to reinvent both science and theorizing itself" (251). A striking claim, accompanied by an equally striking dearth of novel results arising from the allegedly new and revolutionary feminist science (which, of course, should not be taken to be an indictment of feminist critical contributions to science *studies,* as opposed to science per se).

This sort of attitude led to the famous "Sokal affair" (The Editors of Lingua Franca 2000), a practical joke played by physicist Alan Sokal at the expense of the editors of the postmodernist journal *Social Text.* Sokal submitted a paper titled "Toward a Transformative Hermeneutics of Quantum Gravity" (Sokal 1996). Needless to say, there is no such thing as a hermeneutics of quantum gravity, transformative or not. It was a Trojan Horse that Sokal cleverly (although perhaps in an ethically questionable manner) sent to his postmodernist foes. The latter let the horse in and published the paper, with Trojan-like consequences. In a follow up commentary, Sokal exposed their shoddy scholarship and corrupt ideological posturing. Sokal boasted: "When one analyzes [postmodernist] writings [on science], one often finds radical-sounding assertions whose meaning is ambiguous and that can be given two alternative readings: one as interesting, radical, and grossly false; the other as boring and trivially true" (Sokal 1998, 13).

Then again, the blame for the culture wars doesn't lie squarely on the shoulders of humanists. Scientists have engaged in their own overblown rhetoric to aggrandize their doings and unfairly dismiss what they clearly have not read or understood. Interestingly, here the target is, more often than not, philosophy, a field at the crucial juncture—as I see it—between sciences and humanities (after all, philosophy includes the study of logic, and therefore math, and of course science itself began as natural philosophy). Physicist Steven Weinberg (1994) famously wrote an entire essay titled "Against Philosophy" in his *Dreams of a Final Theory*, a sentiment echoed more recently by Stephen Hawking (Hawking and Mlodinow 2012), who began his latest book, *The Grand Design*, by declaring philosophy dead—although he neglected to provide evidence for such a startling conclusion and also undermined his own assertion by immediately proceeding to make what are transparently philosophical arguments about cosmological theories.

Even more recently, yet another physicist, Lawrence Krauss, has abandoned himself (during an interview in the *Atlantic* [Andersen 2012]) to thoughtless dismissals of philosophy:

> Philosophy is a field that, unfortunately, reminds me of that old Woody Allen joke, "Those that can't do, teach, and those that can't teach, teach gym." And the worst part of philosophy is the philosophy of science; the only people, as far as I can tell, that read work by philosophers of science are other philosophers of science. It has no impact on physics whatsoever. . . . they have every right to feel threatened, because science progresses and philosophy doesn't. (Andersen 2012)

To begin with, it is fair to point out that the only people who read works in theoretical physics are theoretical physicists, so by Krauss's own reasoning, both fields are largely irrelevant to everybody else (they aren't, of course). Second, Krauss (and Weinberg and Hawking) does not seem to get that the business of philosophy (and philosophy of science, in particular) is not to solve scientific problems. We've got science for that. To see how absurd Krauss's complaint is just think of what it would sound like if he had said that historians of science haven't solved a single puzzle in theoretical physics. That's because historians do history, not science. When was the last time a theoretical physicist solved a problem in history?

What is going on here? In the influential *Darwin's Dangerous Idea*, philosopher Daniel Dennett (1996) observes: "There is no such thing as philosophy-free science; there is only science whose philosophical baggage is taken on board without examination" (21). Put otherwise, it turns out that people like Weinberg, Hawking, and Krauss need philosophy as a background condition for what they do, but apparently don't even realize it.

E.O. Wilson's Quest for "Consilience"

Perhaps the most ambitious contemporary attempt at restructuring the relationship between the sciences and the humanities has been made by biologist E.O. Wilson (1998) in his book *Consilience: The Unity of Knowledge.* Wilson has recently updated his thinking on specific aspects of his program in articles published in *Harvard Magazine* (2012b) and in the *New York Times* (2012a). It may not come as a surprise that, just as Plato thought that philosophy is the highest calling, Wilson thinks science is where all the action really is (particularly biology, although there is some lip service to physics). Still, it will pay to examine what he and others are proposing in some detail, in part because of the amount of debate it has generated and in part because it is a very good example of what I think is problematic about the unfortunately still ongoing science wars.

Wilson claims we can engage in a process of "consilience" that leads to an intellectually and aesthetically satisfactory unity of knowledge. Here is how he defines two versions of consilience (Wilson 1998): "To dissect a phenomenon into its elements . . . is consilience by induction. To reconstitute it, and especially to predict with knowledge gained by reduction how nature assembled it in the first place, is consilience by synthesis" (74). He is actually referring to a rather standard approach in the natural sciences, which goes back to Descartes' *Meditations* (1996 [1641]): to understand a complex problem, break it down into smaller chunks, get a grasp on them, and then put the whole back together piecemeal. It is the sort of reductionist route to knowledge that has been highly successful in fundamental physics, and has had more limited success in biology (Brigandt 2012) and other natural sciences.

The image that comes to mind here is of a downward spiral where complex aspects of human culture—from music to literature to the performing arts—are understood in terms of social sciences first (sociology, psychology), and then more mechanistically by the biological sciences (neurobiology, evolutionary biology), and finally, possibly, reduced to physics itself. After all, everything is made of quarks (or strings), isn't it?

Before we explore the issue of reductionism in a bit more depth, allow me a small digression on the more commonly accepted meaning of the word *consilience*, which actually Wilson hijacked with little regard toward the field that originated it and still uses it widely: philosophy (although he does cite its historical origin). As it turns out, the term *consilience* was coined by William Whewell (1847) in the mid 19th century (he was also the one who, somewhat ironically, coined the word *scientist*, by analogy with *artist*). Whewell (1847) was involved in a long-standing dispute with John Stuart Mill (2002 [1843]) on the use of inductive reasoning in science, and he proposed that scientific conclusions are arrived at by a convergence of inductions from different classes of facts. Consilience in this context is properly also referred to as inference to the best explanation (and, more bizarrely, as "abduction" [Douven 2011]), indicating it is a method to achieve reasonable (if tentative) conclusions concerning a particular problem, informed by empirical evidence. Both science and common sense work by deploying consilience continuously in this sense. As I'm sure the attentive reader will have noticed, this has to do only superficially with Wilson-type consilience—which is, at its core, a reductionist program. One can see why a philosopher would consider the latter to be a misappropriation or distortion of the former.

Be that as it may, let us go back to Wilson's program of reduction of the humanities to the natural (and eventually physical) sciences. Before we can see where Wilson (and others who followed him more recently) goes wrong, we need to make a distinction between two meanings of reductionism: ontological (that has to do with what exists) and epistemic (that has to do with what we know). Clearly, there are things that exist but we may not know about (how many moons orbit around the fifth planet of the Alpha Centauri system?), whereas the converse is not true: it doesn't make sense to claim knowledge of things that don't exist—except for fictional characters, which "exist" in a highly debated sense of the term of interest to philosophers of language (Carter 1980). Consequently, ontological reductionism is the idea that the bottom level of reality (say, quarks, or strings, or whatever) is causally sufficient to account for everything else (atoms, cells, you and me, planets, galaxies, and so forth; however, see Ladyman and Ross [2009] for a challenge to the whole idea of "a bottom level of reality"). Epistemic reductionism, instead, claims that knowledge of the bottom level is *in principle* sufficient to reconstruct knowledge of everything else, although most epistemic reductionists, when pressed, immediately admit we will never, for practical (and *only* practical, they say) reasons, be able to derive a quantum mechanical theory of Shakespeare's sonnets.

The notion of ontological reductionism is widely accepted in physics and in certain philosophical quarters, although there really isn't any compelling evidence one way or the other (it's a purely metaphysical position). Truth be said, we don't know whether the laws that control the behavior of quarks simply scale up to the level of societies and galaxies or whether, instead, there are so-called emergent properties (in either the weak or strong sense of the term [O'Connor 2012])—in other words, novel behaviors that are characteristic of complex systems and that are not directly reducible to lower ontological levels. I am, therefore, agnostic about ontological reductionism. Fortunately, it turns out that for the purposes of this discussion it doesn't really matter whether you accept or reject ontological reductionism, because the real game lies in the other camp—that of epistemic reductionism.

Any strong version of epistemic reductionism is obviously false. We do not have—nor will likely ever have—a fundamental physical theory of economies or of human psychology. Even if possible in principle (as ontological reductionists claim), such a theory would be too complicated to be computable or simply to be understood by human beings. It is the impossibility of strong epistemic reductionism that, at the very minimum, puts a significant constraint on any Wilson-type consilience. Sure, chemistry may have become a branch of physics (an example of successful reductionism, although even this is actually questionable [Weisberg et al. 2011]), and neurobiology certainly informs psychology. But given that not even the most ardent physicist would attempt to produce an explanation of, say, ecosystems in terms of subatomic particles, clearly there are (epistemic) limits to the reductionist program. The sensible questions, then, are: How far can we push the program? And what do we gain from it?

Why Wilson-Type Consilience Does Not Work

I am going to begin answering these questions by discussing some example of why Wilson-type consilience does not work, pointing out ways in which the search for the "unity of knowledge" appears to be misguided. Let's start in what is perhaps the obvious place: if culture has to be understood in terms of biology, then genes must have quite a bit to do with it. Wilson, however, is too sophisticated to fall into straightforward genetic determinism, so instead he tells us "[g]enes prescribe epigenetic rules, which are the regularities of sensory perception and mental development that animate and channel the acquisition of culture" (Wilson 1998, 171), except this is highly vacuous talk. I have worked on epigenetics (Bossdorf et al. 2008; Richards et al. 2010), the complex molecular processes that mediate the effects of genes during plant and animal development, and I know that biologists don't know what "epigenetic rules" are, how to quantify them, or how to study them. Until further notice, epigenetic rules are a blind alley that does not lead from biology to culture.

Wilson's next choice is to invoke Richard Dawkins and his idea of "memes," or units of cultural evolution (Dawkins 1976; see also Blackmore 1999). You see, if culture is made of discrete units that can replicate and spread in the environment provided by human society, then perhaps there is a way to bring evolutionary theory directly to bear on culture: instead of genes (or epigenes), we apply Darwinian principles to memes.[1]

Except that memetics is in big trouble as a scientific research program. The *Journal of Memetics,* devoted to publishing technical papers in the field, closed a few years ago for lack of submissions. And noted scientists and philosophers have cast serious doubt on the coherence, or at least usefulness, of the very concept of memes. Here is, for instance, my evolutionary biology colleague Jerry Coyne (1999): "[Memetics is] completely tautological, unable to explain why a meme spreads except by asserting, post facto, that it had qualities enabling it to spread" (768). We don't know how to define memes in a way that is operationally useful to the practicing scientist, we don't know why some memes are successful and others not, and we have no clue regarding the physical substrate, if any, of which memes are made. Memes, as it turns out, are an intriguing metaphor, but they do not come close to providing us with even a sketch of a theory of cultural evolution (Burman 2012).

None of this, of course, is to be taken to say that biology is irrelevant to human culture. That would simply be bizarre, *pace* the musings of radical postmodernists. Human beings are biological and physical entities, so what we do and how we do it is constrained and channeled by both the laws of physics (nope, you can't fly out of the window unaided by any type of machinery) and the principles of biology (yes, a lot of what you do in life will have, directly or indirectly, to do with food, sex, and social status). But humanity has also found cultural means to exploit both physics and biology in novel ways. We have built airplanes to fly despite the limitations imposed on our bodies by gravity, and we have invented endlessly fascinating variations on those basic biological themes of love and power, from Shakespeare's tragedies to Picasso's paintings. I'll get back to how I think biology and other sciences do fit with the humanities in a bit, but we still need to look at other problems for Wilson-style consilience and for any similarly totalizing project about human knowledge.

If we take the idea of unity of knowledge seriously, then we better come up with ways of integrating mathematics and logic into our picture as well, and that turns out to be more difficult to do than one might imagine. Wilson (1998) is keen on the role of these disciplines in his grand plan:

> "The dream of objective truth peaked . . . with the formulation of logical positivism . . . that attempted to define the essence of scientific statements by means of logic and the analysis of language. . . . Because of its effectiveness in the natural sciences, mathematics seems to point arrow-like toward the ultimate goal of objective truth. (69)

There are two major obstacles here: one deals with the nature of scientific and mathematical–logical knowledge (Horsten 2012), the other with the pretty well established fact, by now, that human beings simply cannot be in the business of "ultimate objective truth" (Giere 2010).

So, what is the nature of scientific and mathematical–logical knowledge? As it turns out, they are—obviously, I think—quite different. Consider what counts as a "fact" in science—for instance, the statement that there are four natural satellites of Jupiter that can be seen through small telescopes from Earth. These satellites were discovered by Galileo Galilei during the 17th century and represented the first example of a solarlike system within our own sun-centered one. Indeed, Galilei used this as a major reason to convince others to take seriously the then-new and highly controversial Copernican theory.

In contrast, consider a mathematical "fact," such as the demonstration of the Pythagorean Theorem. This looks nothing like a fact as it is understood in the natural sciences, which I suggest implies that the word *knowledge* (of facts), if used in both cases, actually indicates a heterogeneous category. If we don't take this sort of consideration onboard, we simply fail to understand that "unifying knowledge" in this case is an empty aim, unless by unifying we simply mean that we recognize that we have natural sciences over here and math over there, and the latter is often useful (for not at all clear reasons, by the way [Linnebo 2011]) to the former.

The same goes for logic. Logical "facts" look like, say, a truth table that tells you the conditions under which particular combinations of premises yield true or false conclusions when the rules of deduction are applied to them. If anything, this sort of fact is much more akin to mathematical facts than to facts from the natural sciences, which is why some philosophers think of math as a type of logical system, but nobody refers seriously either to math or logic as "sciences." The point is that if consilience means to reduce knowledge in one field to knowledge in another, more fundamental one, we already run into trouble as soon as we attempt consilience between natural sciences and logic–math—and that ought to be the easy part! Knowledge being a heterogeneous category, to apply the word without qualification across fields leads to what philosophers call *category mistakes*, like asking what is the smell of triangles.

Let's consider yet another type of fact, more germane to the difficulties intrinsic in reducing the humanities to the sciences. I happen to have a very strong conviction that the music of Ludwig van Beethoven is immeasurably superior to that of Britney Spears (strictly speaking, of course, the comparison itself is dubious, because Beethoven was a composer, whereas someone else writes Ms. Spears' songs. Still, the point of the analogy will be obvious). To me, that's an aesthetic fact. I hope it's also very clear that this is a "fact" (based on my "knowledge" of music) that has a very different structure and content from both logical–mathematical and natural scientific facts. Indeed, it isn't a fact at all, it's actually an aesthetic *judgment* to which I have a strong emotional attachment. Now, I do not doubt that my ability to make aesthetic judgments of any sort is in part influenced by the kind of biological being that I am. For instance, I need to have a particular type of auditory system even to hear either Beethoven or Spears, which surely explains why musicians don't produce pieces at sound frequencies that cannot be heard by their fellow humans. But it seems very hard to argue that

my judgment about Beethoven versus Spears is not primarily the result of my culture and how it shaped my psychology during my upbringing. People in different times and cultures, or with different temperaments, have disagreed and will disagree with my take—and they might feel just as strongly about their judgment as I do about mine (of course, they would be "wrong").

Examples like these ought to make clear that there are plenty of aspects of human culture in which the very notion of objective truth is a category mistake. But things are even worse for Wilson and his followers, because we have very good reasons to believe their goal of objective truth cannot be achieved even by the natural sciences, in fact, not even by logic and math, those "arrowlike" pointers Wilson mentioned earlier!

The Failure of Positivism and the Search for Objective Truth

Wilson is fond of two philosophical movements: the Enlightenment of the 18th century, and the already encountered logical positivism of the early 20th century. Of the latter he says (Wilson, 1998): "Its failure, or put more generously, its shortcoming, was caused by ignorance of how the brain works. That in my opinion is the whole story" (69). That is a very strange opinion. To begin with, it is hard to see what neuroscience has to do with any of this at all. Neuroscience may (and does) shed light on how people acquire and process knowledge of the external world, but it has no relevance to settling the sort of philosophical questions about truth raised by the logical positivists. Indeed, logical positivism is beyond rehabilitation, philosophically speaking, and Wilson's prediction of its resurgence would be stunning news to philosophers, who have mounted a number of devastating objections to the concept of objective knowledge put forth by the logical positivists and their American counterparts, the logical empiricists (Creath 2011). To remind ourselves of what this is about, remember that a fundamental idea put forth by logical positivists was the so-called verifiability principle, according to which the only meaningful sentences are those whose truth can be verified, either empirically or mathematically. The positivists used their principle to argue that metaphysical notions, say, are not just wrong or unverifiable, they are literally meaningless. Critics of logical positivism have pointed out a variety of problems with this, the most embarrassing of which, of course, is that there is no way to apply the verifiability principle to the principle itself (interestingly, an objection following the same fundamental logic applies to any form of strong cultural relativism, such as some versions of postmodernism). That is, the verifiability principle is nonsense by its own standard of meaningful discourse. (This said, Ladyman and Ross [2009], have begun to develop a form of "neo-positivism" that relies on a Peirce-inspired, pragmatic version of verificationism.)

Regardless of the specific failure of positivism, though, how are we doing in the millennia-long quest for absolute and objective truth, which many philosophers and scientists have been dreaming of? Not so well, it seems, largely because of the devastating contributions of a few philosophers and logicians, particularly David Hume, Bertrand Russell, and Kurt Gödel.

Hume (1995 [1748]) famously formulated what is now known as the problem of induction. He noted that both in science and in every day experience we use a type of reasoning that philosophers call *induction*. It consists of generalizing from examples, and of exercising Whewell–style consilience.[2] Hume also pointed out that we do not seem to have a logical justification for the inductive process itself: why do we think that inductive reasoning is a good approach to solve everyday as well as scientific problems? The best and immediately intuitive answer is that it works, or at least it has worked so far. But to say so is to deploy inductive reasoning to justify inductive reasoning, which is obviously circular. Plenty of philosophers have written about the problem of induction (see, particularly, Okasha [2001]), but it seems pretty resistant to solutions. We do not have an independent, rational justification for the most common type of reasoning used by laypeople and professional scientists alike. Hume didn't conclude from this that we should therefore all quit and go home. Indeed, we don't have any choice but to keep using induction. But it ought to be a sobering thought that all our empirical knowledge is not based on any solid foundation other than "it works."

What about math and logic? As is well known, at the beginning of the 20th century a number of logicians, mathematicians, and philosophers of mathematics were engaged in a quest to establish firm logical foundations for mathematics and similar formal systems. The most famous such an attempt was carried out by Alfred North Whitehead and Bertrand Russell (2010 [1927]), and resulted in their ponderous *Principia Mathematica*, arguably one of the most difficult books to read of all times. They failed, and a few years later, logician Kurt Gödel (1962 [1931)] helped explain why. His two "incompleteness theorems" proved—logically!—that any sufficiently (and interestingly) complex mathematical or logical system will contain truths that cannot be proved from within that system. Russell realized that this was a blow to his enterprise, and that we have to be content with only partial logical justification even in mathematics. If we add to

Gödel's results the well-known fact that logical proofs and mathematical theorems have to start from assumptions (or axioms) that are themselves unprovable (or, in the case of some deductive reasoning such as syllogisms, are derived from empirical observations and generalizations—in other words, by induction!), it seems the quest for true and objective knowledge is quickly revealed as a mirage. (There are many more reasons from epistemology and philosophy of science, but I think we don't need any additional nails in that particular coffin.)

Now, you will remember that Wilson stated his belief that the "shortcomings" of logical positivism, and hence the failure of our search for objective truth, will be reversed when we overcome our "ignorance of how the brain works." It should be clear by now why that will never be the case, but let me be explicit just in case. Neurobiology is a fascinating enterprise, but it has nothing whatsoever to tell us about epistemology and truth. Perhaps a simple example will clarify what I mean. Let's say a neuroscientist is going to run a functional magnetic resonance image of your brain while you are engaged in writing out a proof of Pythagoras' Theorem. The image will tell you something (interesting, for sure) about which areas of your brain are involved in what you are doing—in other words, it will pinpoint the neural correlates of your mathematical reasoning. What no functional magnetic resonance image will ever be able to tell you is whether you got the proof right. And that really is the whole story, in my opinion.

The *Pars Construens* of the Consilience Project

Enough with the demolition project. Let us start putting the pieces back together and see if we can reconstruct something like Wilson's consilience, but in a more reasonable manner. My colleague David Sloan Wilson (no relation to E.O.), after an animated discussion we had on such matters at the consilience conference, calls this "congruence" of science and humanities. As you will see, this is a much more appropriate term for the project of building a sensible bridge crossing Snow's two-culture divide.

Consider visual art. Its history comprises prehistoric cave paintings, Michelangelo, Picasso, and contemporary abstract art, to name but a few. It is reasonable to think that science—perhaps a combination of evolutionary biology and cognitive science—can tell us something relevant and interesting about why our ancestors started painting to begin with, as well as why we like certain types of patterns produced by artists (for example, we like symmetrical figures and some repetitive patterns of a certain degree of complexity). But these sorts of explanations massively underdetermine the bewildering variety of ways of doing visual arts, both across centuries and across cultures. Picasso's cubism is not about symmetry, for instance; indeed, it's about *breaking* symmetry. And it is hard to imagine an explanation of the rise of, say, the Impressionist movement without invoking the specific cultural circumstances of late-19th-century France, and possibly even the individual biographies and psychologies of artists such as Monet, Cézanne, Degas, and company.

Or consider, instead, math. It is likely that our ability to learn how to count and do simple arithmetic gave us an evolutionary advantage, and was therefore possibly the result of natural selection. (Notice, however, that this is a plausibility argument only, because we do not have access to the kind of empirical evidence that would be necessary to actually test the hypothesis [Kaplan 2002].) But what on earth is the possible adaptive value of highly abstract mathematics? Why would evolution produce brains like Andrew Wiles's (1995), capable of solving the infamous Fermat's last theorem? Biology sets the background conditions for such feats of human ingenuity, because a brain of a particular type is necessary to accomplish them. But biology by itself has little else to say about how some human cultures took a historical path that ended up producing a small group of stereotypically socially awkward people who devote their lives to solving abstruse mathematical problems.

Or, finally, take morality, perhaps the most defining characteristic of what it means to be human. Much has been written on the evolutionary origins of morality, and I think people have proposed many good and plausible ideas (again, difficult to test empirically) about it. Very likely our moral sense originated in the context of social life as intelligent primates, a supposition validated by the fact that other social primates also show behaviors consistent with the basic building blocks of morality (such as fairness toward other members of the group, even nonkin [de Waal 2009]). But it is a very long way from that to Aristotle's *Nicomachean Ethics,* Immanuel Kant's categorical imperative, and Jeremy Bentham's and John Stuart Mill's utilitarianism. These works and concepts were, again, possible because we are biological beings of a certain kind (social, intelligent), but we need to take cultural history, psychology, and philosophy seriously to account for them and for their (huge) influence on the way we think, live our lives, and structure our societies.

Wilson–style reductionist consilience is predicated on the explicit assumption that there is such a thing as human knowledge as a unifiable category, an assumption that I challenged earlier. This disunity of knowledge may account for the fact that we have different disciplines studying different aspects of the human quest for understanding. Although for Wilson disciplinary boundaries are arbitrary accidents of history that need to be eliminated, there are other ways of accounting for why we speak of philosophy, literature, psychology, biology, physics, and so on, in the way we do. An intriguing suggestion has been made, in different contexts, by linguist Noam Chomsky (1975) and philosopher Colin McGinn (1991).

The basic idea is to take seriously the fact that human brains evolved to solve savanna-induced problems during the Pleistocene, not to engage in the ultimate quest for the nature of reality. From this perspective, it is delightful that we do discover as much as science allows us to discover, or to ponder as much as philosophy makes it possible for us to ponder. But we know that there are cognitive limits to the power of the human mind (just for fun, try to memorize a sequence of a million digits), and perhaps at least some of the disciplinary boundaries that have evolved over the centuries simply reflect our epistemic limits. Seen this way, the differences between philosophy, biology, physics, the social sciences, and so on, may not be the result of arbitrary caprice of academic administrators and faculty, but rather may reflect a more or less natural way in which human beings understand the world and their role in it. There may be better ways to organize our knowledge, in some absolute sense, but likely what we have come up with is something that works well for us as biological–cultural beings of a certain type and with a certain history.

To put it as evolutionary biologist Allen Orr (1998) did in his review of Wilson's book:

> The real reason Wilson favors his consilient scenario isn't because he finds it more plausible but because he finds it more attractive. For as he admits near the start of his book, consilience isn't science, it is a philosophy, a

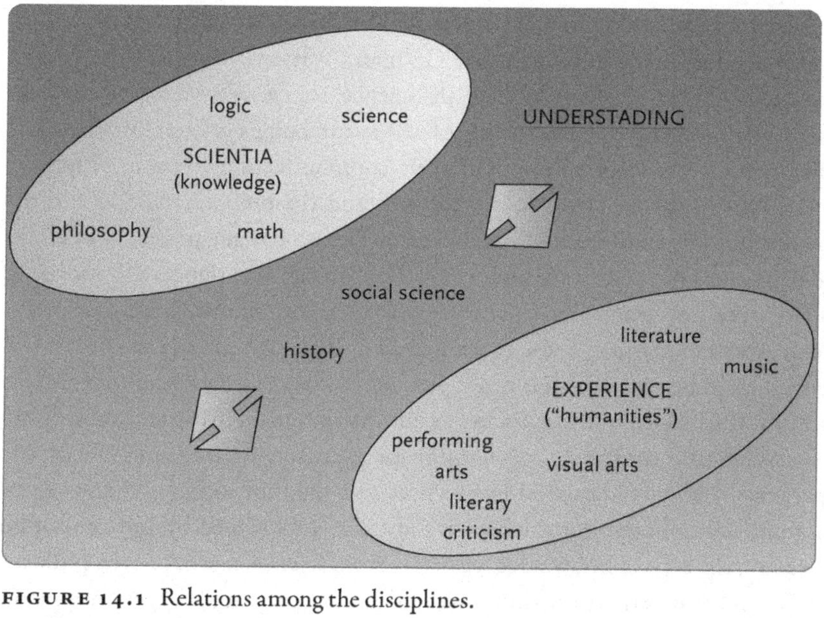

FIGURE 14.1 Relations among the disciplines.

> metaphysical view that he obviously finds both beautiful and deeply satisfying. The irony, of course, is that Wilson's own science of evolution gives every reason for questioning this metaphysic, every reason, that is, for doubting whether our brains—jury-rigged and riddled with blind spots—are the stuff from which certain knowledge and seamless consilience can be obtained. (http://bostonreview.net/archives/BR23.5/Orr.html, accessed on 6 January 2016)

This isn't a council for despair, or a suggestion to give up. It's simply a bit more humble of a take on human knowledge, and one that is ironically more in synch with what the natural sciences tell us about being human than Wilson and colleagues' somewhat quixotic quest for objective truth or ultimate knowledge.

A Different Model for Consilience

Where does all of this leave us? If I am right, we are looking at a more nuanced set of relationships between the natural, physical, and social sciences and the humanities, something along the lines of that shown in Figure 14.1.

There are three key concepts in the diagram: knowledge, experience, and understanding. Although I don't necessarily see any sharp boundaries here, *knowledge* refers to the sort of thing we acquire by rational thinking and that, when possible and appropriate, is evidence based. This encompasses science *sensu stricto,* logic, mathematics, and philosophy. It is what the Latin used to call *scientia,* or knowledge in the broad sense (akin to the German "Wissenschaft" and the French "science," and broader than the English "science"). *Experience* refers to the first-person qualia that are characteristic of self-aware beings capable of reflection, their immediate interpretation, and their communication to other self-aware beings. As such, it encompasses all the arts and the humanities. You can tell someone how his enjoyment of Beethoven or Picasso is made possible by certain brain circuits, but you can't actually share that private experience, the outline of which can be communicated to other people only through language that is outside the scope of *scientia.* These two domains overlap and talk to each other, and the mediation between them is made possible by the study of disciplines such as history and the social sciences, the very same disciplines often disparaged as "soft" sciences that connect the logical–naturalistic approach to being human with the experience itself as interpreted by the arts and the humanities.[3] The whole of it amounts to *understanding,* which is most certainly helped by, but cannot be limited to, the type of knowledge we derive from science, math, or even philosophy. This is because there are things we understand because we experience them, regardless of how well the hard sciences can explain their physical bases. As philosopher Thomas Nagel (1974) famously put it, there is something that it is like to be a bat, and the only way to find out is to *be* a bat, quite independently of how well science can account for the mechanics of echolocation.

None of this, incidentally, should be interpreted as giving any consolation to mystical mumbo jumbo about going "beyond science." There is nothing beyond science, but there is important stuff *before* (or on the side of) science; there are human emotions, expressed by literature, music, and the visual arts. The best understanding of the whole shebang that humanity may hope to achieve, then, goes through a continuous—and respectful—dialogue between the various disciplines that contribute to human knowledge and experience, leading to understanding in the broadest sense. And we would be much poorer emotionally and intellectually if we ignored the sciences or tried too hard to reduce the humanities.

Notes

1. Wilson uses a slightly different definition of memes from the original Dawkinsian one: "The definition of meme I suggest is nevertheless more focused and somewhat different from that of Dawkins. It is the one posed by the theoretical biologist Charles J. Lumsden and myself in 1981, when we outlined the first full theory of gene-culture coevolution. We recommended that the unit of culture—now called meme—be the same as the node of semantic memory and its correlates in brain activity" (Wilson 1998, 136). Because the notion of a "node of semantic memory and its correlates in brain activity" is extremely vague, things don't actually get much better for memetics by framing the issue this way.
2. Some logicians consider inference to the best explanation to be different in kind from standard induction. This debate does not change the gist of my argument here, because Hume's problem applies both to standard induction as well as to abduction.
3. To be clear, I am not saying, for instance, that historians use literature to bridge a gap toward science. I mean to say that understanding broadly construed has to make use of the so-called soft sciences to help us—humanity in general—bridge the gap between technical knowledge and personal experience.

References

Andersen, Ross. 2012. Has physics made philosophy and religion obsolete? *Atlantic* April 23, http://www.theatlantic.com/technology/archive/2012/04/has-physics-made-philosophy-and-religion-obsolete/256203/ (accessed on 6 January 2016).

Blackmore, Susan. 1999. *The meme machine.* Oxford: Oxford University Press.

Bossdorf, Oliver, Christina L. Richards, and Massimo Pigliucci. 2008. Epigenetics for ecologists. *Ecology Letters* 11: 106–115.

Brigandt, Ingo. 2012. Reductionism in biology. In *Stanford encyclopedia of philosophy.* [Online].

Burman, Jeremy T. 2012. The misunderstanding of memes: Biography of an unscientific project, 1976 to 1999. *Perspectives on Science* 20: 75–104.

Carter, William R. 1980. Do creatures of fiction exist? *Philosophical Studies* 38: 205–215.

Chappell, Timothy. 2009. Plato on knowledge in the Theaetetus. In *Stanford encyclopedia of philosophy.* http://plato.stanford.edu/entries/plato-theaetetus/ [Online].

Chomsky, Noam. 1975. *Reflections on language.* New York: Pantheon Books.

Coyne, Jerry A. 1999. The self-centred meme. *Nature* 398: 767–768.

Creath, Richard. 2011. Logical empiricism. In *Stanford encyclopedia of philosophy.* http://plato.stanford.edu/entries/logical-empiricism/ [Online].

Dawkins, Richard. 1976. *The selfish gene.* Oxford: Oxford University Press.

Dennett, Daniel. 1996. *Darwin's dangerous idea: Evolution and the meanings of life.* New York: Simon and Schuster.

Descartes, René. 1996 [1641]. *Meditations on first philosophy.* Cambridge: Cambridge University Press.

de Waal, Frans. 2009. *Primates and philosophers: How morality evolved.* Princeton: Princeton University Press.

Douven, Igor. 2011. Abduction. In *Stanford encyclopedia of philosophy.* http://plato.stanford.edu/entries/abduction/ [Online].

Giere, Ron N. 2010. *Scientific perspectivism.* Chicago: University of Chicago Press.

Gödel, Kurt. 1962 [1931]. *On formally undecidable propositions of principia mathematica and related systems.* Mineola, NY: Dover.

Harding, Sandra. 1986. *The Science Question in Feminism: Industrial Policy in Europe.* Ithaca, NY: Cornell University Press.

Hawking, Stephen, and Leonard Mlodinow. 2012. *The grand design.* New York: Bantam.

Horsten, Leon. 2012. Philosophy of mathematics. In *Stanford encyclopedia of philosophy.* http://plato.stanford.edu/entries/philosophy-mathematics/ [Online].

Hume, David. 1995 [1748]. *An inquiry concerning human understanding.* Upper Saddle River, NJ: Prentice Hall.

Kaplan, Jonathan M. 2002. Historical evidence and human adaptations. *Philosophy of Science* 69: S294–S304.

Labinger, Jay A., and Henry Collins, eds. 2001. *The one culture? A conversation about science.* Chicago: University of Chicago Press.

Ladyman, James, and Don Ross. 2009. *Every thing must go: Metaphysics naturalized.* Oxford: Oxford University Press.

Linnebo, Øystein. 2011. Platonism in the philosophy of mathematics. In *Stanford encyclopedia of philosophy.* http://plato.stanford.edu/entries/platonism-mathematics/ [Online].

McGinn, Colin. 1991. *The problem of consciousness: Essays towards a resolution.* Hoboken, NJ: Wiley-Blackwell.

Mill, John S. 2002 [1843]. *System of logic, ratiocinative and inductive.* Honolulu: University Press of the Pacific.

Nagel, Thomas. 1974. What is it like to be a bat? *The Philosophical Review* 83: 435–450.

O'Connor, Timothy. 2012. Emergent properties. In *Stanford encyclopedia of philosophy.* http://plato.stanford.edu/entries/properties-emergent/ [Online].

Okasha, Samir. 2001. Hume and induction. *Philosophical Quarterly* 51: 307–327.

Orr, Allen. 1998. The big picture: E.O. Wilson's new book is ambitious, vague, and philosophically naïve, *Boston Review* October/November.

Richards, Christina L., Oliver Bossdorf, and Massimo Pigliucci. 2010. What role does heritable epigenetic variation play in phenotypic evolution? *BioScience* 60: 232–237.

Snow, Charles P. 2012 [1959]. *The two cultures.* Cambridge: Cambridge University Press.

Sokal, Alan. 1996. Toward a transformative hermeneutics of quantum gravity. *Social Text* 14: 217–252.

Sokal, Alan. 1998. What the *Social Text* affair does and does not prove. In N. Koertge (ed.), *A House Built on Sand: Exposing Postmodernist Myths About Science.* Oxford: Oxford University Press.

The Editors of Lingua Franca. 2000. *The Sokal hoax: The sham that shook the academy.* Winnipeg, Manitoba: Bison.

Weinberg, Steven. 1994. Against philosophy. In Steven Weinberg (ed.), *Dreams of a final theory: The scientist's search for the ultimate laws of nature*, 166–190. New York: Vintage.

Weisberg, Michael, Paul Needham, and Robin Hendry. 2011. Philosophy of chemistry. In *Stanford encyclopedia of philosophy.* http://plato.stanford.edu/entries/chemistry/ [Online].

Whewell, William. 1847. *The philosophy of the inductive sciences, founded upon their history.* Mira Loma, CA: Parker.

Whitehead, Alfred N., and Bertrand Russell. 2010 [1927]. *Principia mathematica.* Charleston, SC: Nabu Press.

Wiles, Andrew. 1995. Modular elliptic curves and Fermat's last theorem. *Annals of Mathematics* 141: 443–551.

Wilson, Edward O. 1998. *Consilience: The unity of knowledge.* New York: Vintage.

Wilson, Edward O. 2012a. Evolution and our inner conflict. *New York Times,* June 24. http://opinionator.blogs.nytimes.com/2012/06/24/evolution-and-our-inner-conflict/ (accessed on 6 January 2016).

Wilson, Edward O. 2012b. On the origins of the arts. *Harvard Magazine* May/June. http://harvardmagazine.com/2012/05/on-the-origins-of-the-arts (accessed on 6 January 2016).

AFTERWORD

David Sloan Wilson

This volume is a milestone on a long trek toward a view of human nature that does justice to both science and the humanities. This particular milestone shows we have come a long way, and also that we have a long way to go. The chapters are as interesting for their discordances as for the unity the authors are striving to achieve.

The very word *consilience* requires clarification, as Massimo Pigliucci highlights in Chapter 14. At least two major meanings are on display in this volume that need to be distinguished. The first is physical reductionism—the claim that "all complex phenomena can be reduced to relations among simpler elements," as Joseph Carroll puts it in his introduction to this volume. This claim can be discomforting to scholars in the humanities because it makes it appear their knowledge will be subsumed by the sciences. As Alice Dreger puts it in her foreword to the volume, "To tell many humanists that the subjects of our attractions are 'reducible' to biology (and then to chemistry, and then to physics) is, I think, heard as telling us that we don't know our own work." Why shouldn't humanists be discomfited when the first sentence of E.O. Wilson's contribution to the volume (Chapter 1) is "The meaning of humanity is too important a subject to leave to the humanities?"

The distinction between proximate and ultimate causation in evolutionary theory (Mayr 1961) shows why humanists need not be threatened by physical reductionism. Everything that evolves requires two complementary explanations: why it evolved compared with many other traits that could have evolved (ultimate causation) and how it exists in a physical sense (proximate causation). The ultimate explanation of a trait cannot be reduced in the same way as the proximate explanation. Apple blossoms bloom in spring because those that bloomed earlier were nipped by frost and those that bloomed later didn't have time to develop their fruits (ultimate causation). This statement can be made without knowing anything about the physical makeup of apple trees (as long as the physical makeup results in heritable variation). Apple blossoms also bloom in spring because of a physiological mechanism that can be understood reductionistically—that's what molecular biology is all about.

Knowledge of proximate causation can only complement and never substitute for knowledge of ultimate causation. The most consilient research programs in evolutionary biology study both in conjunction with each other, along with their temporal components, development (the temporal component of proximate causation) and phylogeny (the temporal component of ultimate causation [Tinbergen 1963]). This fully rounded view of evolution provides much more comfort to humanists than one that focuses entirely on physical reductionism. A number of authors in this volume make this point, which I am pleased to showcase in my afterword.

The second meaning of consilience on display in *Darwin's Bridge* is a call for consistency among academic disciplines. Physics, chemistry, and biology are arguably more consistent with each other than the human-related sciences and the humanities. As an example, the so-called orthodox model in economics makes certain assumptions about human preferences and abilities that are at odds with the facts of human psychology. They also make assumptions about economic systems being at equilibrium that are at odds with real economic systems. This lack of consistency is something that needs to be addressed, regardless of what one might think of physical reductionism (see Chapter 3 and Chapter 5 for more on economics from an evolutionary perspective). There is widespread agreement among the authors of this volume on the need for consistency among disciplines, in the humanities and behavioral sciences no less than biology, chemistry, and physics. What would be a cogent argument *against* consistency?

As we work toward consistency, evolutionary theory needs the humanities and human-related sciences more than most evolutionary biologists are prepared to admit. Evolution requires three ingredients: variation, selection, and heredity, defined as a resemblance between parents and offspring. The study of heredity became focused almost entirely on genes during the 20th century, so much so that many evolutionary biologists define evolution as genetic evolution. This position is indefensible because any other mechanism that creates a resemblance between parents and offspring is equally evolutionary. Other mechanisms of inheritance include epigenetics (which involves changes in gene expression rather than gene frequency), forms of social learning found in many species, and forms of symbolic thought that are distinctively human (Jablonka and Lamb 2006). Epigenetics is a biological topic, but social learning and human symbolic thought are the province of the human-related sciences and humanities, making them essential disciplines for expanding core evolutionary science beyond its nearly exclusive focus on genetic evolution.

When the study of evolution becomes focused on variation, selection, and heredity rather than genes per se, then we can proceed to study human cultural diversity in the same way that we currently study biological diversity. The study of proximate mechanisms will add to humanistic inquiry rather than appearing to threaten it. That is the destination toward which we are heading, at least in my opinion, and *Darwin's Bridge* provides a way for the reader to join the authors on their way to the next milestone.

References

Jablonka, Eva, and Marion J. Lamb. 2006. *Evolution in four dimension: Genetic, epigenetic, behavioral, and symbolic variation in the history of life*. Cambridge, MA: MIT Press.

Mayr, Ernst 1961. Cause and effect in biology. *Science* 134: 1501–1506.

Tinbergen, Niko 1963. On aims and methods of ethology. *Zeitschrift für Tierpsychologie* 20: 410–433.

AFTERWORD

Jonathan Gottschall

About 15,000 years ago in the foothills of the Pyrenees, a man stood in a river that flowed from a cave. The man waded into the cave with his bag and torch held high, water soaking his beard. After a few minutes he clambered onto a thin gravel beach, where he stood for time, chattering in a fragile bubble of torchlight.

Holding tight to the wall and crunching the gravel with his feet, he moved on. He looked up to see stone daggers hanging from the ceiling and the shapes of ibexes and bison gouged into the walls. He stopped beneath a shaft that stretched up vertically for 40 feet. The shaft was full of branches that men had forced in cross-wise, and he climbed them like a ladder.

He emerged in a space so tight that even a child would have to stoop. He pressed forward on his hands and knees, the walls tightening around him. At times he could walk. At times he had to slither on his belly, angling to slide his broad shoulders deeper and deeper into the earth. He skirted pitfalls that opened up like deep, hungry throats. He passed the bones of cave bears. He passed into rooms where quartz glittered like stars.

He reached his destination: a chamber shaped like a huge tipped bowl. The man took hollowed rocks from his bag that were full of congealed fat. He lit the wicks that were standing in the fat, and the lamps sputtered and smoked and threw out light. He kneeled to dig and scrape at the floor. He piled and packed clay against a low boulder and began to work it using hands and stones and the horn of an ibex. He shaped two clay bison, a bull rising up to mount a cow, and polished them smooth with puddle water.

This really happened, more or less. Fifteen thousand years ago in France, a person or persons really did swim, climb, slither, and crawl almost a kilometer down into the earth to make art. And we know that his people went there to marvel; you can still see the footprints of men, women, and children running down the cave's hardened mud corridors (Breuil 1979; Begouen et al. 2009).

The 1912 discovery of the clay bison of the Tuc D'Audobert caves was one of many shocking 20th-century discoveries of sophisticated cave art stretching back tens of thousands of years. The discoveries transformed our sense for who our caveman ancestors were. They were not furry, grunting troglodytes. They had artistic souls. They showed us that humans are—by nature, not just by culture—art-making, art-consuming, art-addicted apes.

Prehistoric people lived short, hard, dangerous lives. They faced hostile tribes and dangerous beasts and cruel winters. There were mates to woo, children to feed, rivals to defeat. Why did they brave those caves with their bears and killer mazes and demon-haunted blackness? Why did they go there to paint and to sculpt, and maybe to sing, dance, and tell stories?

No one knows for sure. And here's one reason why: science has mainly ignored the question.

A long time ago, someone proclaimed that art stood beyond the reach of science, and for some reason almost everyone believed it. The humanities and sciences constituted, as Stephen Jay Gould might have proclaimed (see Gould 1997), separate, nonoverlapping magisteria—the tools of the one were radically unsuited to the other (for a critique of this idea, see Gottschall [2008]).

In the humanities, the rightness of this separation is accepted uncritically. Leading thinkers vigilantly patrol the border between the two cultures, ready to sing out "scientism!" should they detect violations of the line (e.g., Menand 2005; Goodheart 2007; for an attempt to rebrand scientism as a good thing, see Pinker [2013]). Scientists don't actively guard the border between the two cultures, but they nonetheless behave as though it is real. How else can we explain science's neglect of the arts? People live in art. We read stories, watch them on TV, and listen to them in our songs. We make paintings and gaze at them on walls. We beautify our homes like bowerbirds adorning nests. We demand beauty in the products we buy, which explains the gleam of our automobiles and the sleek modernist aesthetic of our iPhones. We make our own bodies into art, sculpting them through diet and exercise, festooning them with jewelry and colorful garments, turning our skin into living canvas for the display of tattoos. And so it is the world over. As Denis Dutton (2009) argues in *The Art Instinct*, underneath the brilliant scrim of cultural variation, "all human beings have essentially the same art" (29).

Our curious love affair with art sets our species apart as much as our sapience or our language or our use of tools. And yet we understand so little about it. We don't know why art exists in the first place. We don't know why we crave beauty. We don't know how art produces its effects in our brains—why one arrangement of sound or color pleases but another cloys. We don't know very much about the precursors of art in other species, and we don't know when humans became creatures of art in the first place. (According to one influential theory, art arrived 50,000 years ago in a creative Big Bang [Mithen 1996]. If that's true, how did that happen?) We don't even have a good definition, in truth, for what art is. In short, there is nothing so central to human life that is so incompletely understood.

Darwin's Bridge asks an audacious question: What if the borderline between two cultures of the humanities and sciences has no substance? What if all that's stopping the free flow of concepts, information, and methods is mental and bureaucratic inertia? This is a controversial notion, but the chapters in this book illustrate something that should never have been controversial: a consilient viewpoint that acknowledges the interactive relationships of biology *and* culture in human affairs must be more moderate, holistic, and inclusive than the rigidly reductive and deterministic blank-slate theories that have dominated humanities scholarship for generations. This volume's contributors fully acknowledge the power of acculturation while they resurrect *Homo sapiens*—the mammal, the Old World Catarhine primate, the hominin who has known agriculture and civilization for just 10,000 years.

Darwin's Bridge is part of a new movement by scientists and scholars to bridge the two-culture divide (for an overview of research, see Gottschall [2012]). Neuroscientists can show us what's happening in the brain when we enjoy a song or study a painting. Psychologists are studying the ways novels and TV shows shape our attitudes and behaviors, our politics and our morality. Evolutionary psychologists and literary scholars are teaming up to explore narrative's Darwinian origins. Other literary scholars are developing a "digital humanities," using algorithms to extract some of the information in literature. And the contributors to this volume are bringing the consilience framework to questions ranging from the meaning of life and free will, to the analysis of professional wrestling, horror stories, primitive mark-making, and 19th-century novels.

But consilient work in the humanities is still in its infancy. It is not yet an established research program. If we want better answers to fundamental questions about art—if we want to know why an ancient sculptor made art in the entrails of the earth or why his descendants still care—we must have a lot more of the border-crashing reflected in this volume. Going it alone, humanities scholars can tell intriguing stories about the origins and significance of art, but they don't have the tools to winnow patiently the field of competing ideas. That's what the scientific method is for—separating the stories that are more accurate from the stories that are less accurate. But make no mistake, a consilient study of art will require both the thick, granular expertise of humanities scholars and the clever hypothesis testing of scientists. I'm not calling for a scientific takeover of the arts, and neither are the other contributors to this volume. This is a call for partnership.

This partnership faces great obstacles. There's the unexamined assumption that something in art makes it science-proof. There's a widespread, if usually unspoken, belief that art is just a frill in human life—relatively unimportant compared with the weighty stuff of science. And there's the weird idea that science necessarily destroys the beauty it seeks to explain (as though a learned astronomer really could dull star shine). But the Delphic admonition "know thyself" still rings out as the great prime directive of intellectual inquiry, and there will always be a gaping hole in human self-knowledge until we bring true consilience to the study of art.

References

Begouen, Robert, Carole Fritz, Gilles Tosello, Jean Clottes, Andreas Pastoors, and Francois Faist. 2009. *Le sanctuaire secret des bisons*. Paris: Somogy.

Breuil, Abbé Henri. 1979. *Four hundred centuries of cave art*. New York: Hacker Art Books.

Dutton, Denis. 2009. *The art instinct: Beauty, pleasure, and human evolution*. New York: Bloomsbury Press.

Goodheart, Eugene. 2007. *Darwinian misadventures in the humanities*. Piscataway, NJ: Transaction.

Gottschall, J. 2008. *Literature, science, and a new humanities*. New York: Palgrave Macmillan.

———. *The storytelling animal: How stories make us human*. New York: Houghton Mifflin Harcourt.

Gould, Stephen Jay. 1997. Nonoverlapping magisterial. *Natural History* 106: 16–22.

Menand, Louis. 2005. Dangers within and without. *Profession* 10–17. http://www.jstor.org/stable/25595793?seq=1#page_scan_tab_contents

Mithen, Steven J. 1996. *The prehistory of the mind: A search for the origins of art, religion, and science*. London: Thames and Hudson.

Pinker, Steven. 2013. Science is not your enemy: An impassioned plea to neglected novelists, embattled professors, and tenure-less historians. *The New Republic* August 6. http://www.newrepublic.com/article/114127/science-not-enemy-humanities [Online].

CONTRIBUTORS

Christopher Boehm is a professor of biological sciences and anthropology, and director of the Goodall Research Center at the University of Southern California. His interests include human evolution, political evolution, moral behavior, and development of a hunter-gatherer database. Boehm is a recipient of Simon Guggenheim Fellowship and Santa Fe School of Advanced Research Fellowship. His recent publications include *Moral Origins: The Evolution of Virtue, Altruism and Shame* (Basic Books, 2012) and "Ancestral Hierarchy and Conflict" (*Science*, 2012).

Brian Boyd is university distinguished professor in English, drama, and writing studies at the University of Auckland. Long known as a Nabokov scholar, he has also recently worked on literature, evolution, and cognition, including *On the Origin of Stories: Evolution, Cognition, and Fiction* (Harvard University Press, 2009); *Evolution, Literature, and Film: A Reader*, co-edited with Joseph Carroll and Jonathan Gottschall (Columbia University Press, 2010); and *Why Lyrics Last: Evolution, Cognition, and Shakespeare's Sonnets* (Harvard University Press, 2012). He is currently researching a biography of philosopher Karl Popper.

Joseph Carroll is curators' professor of English at the University of Missouri, St. Louis. His books include *Evolution and Literary Theory* (University of Missouri Press, 1995), *Literary Darwinism* (Routledge, 2004), *Reading Human Nature* (SUNY Press, 2011), and *Graphing Jane Austen* (co-author; Palgrave Macmillan, 2012). He produced an edition of *On the Origin of Species* (Broadview, 2003) and co-edited with Brian Boyd and Jonathan Gottschall *Evolution, Literature, and Film: A Reader* (Columbia University Press, 2010).

Mathias Clasen is an assistant professor in the Department of English at Aarhus University. He specializes in horror fiction from an evolutionary perspective. He has written on postapocalyptic fiction, zombies, vampires, and other supernatural monsters, and he is currently writing a book on horror in fiction, film, and video games.

Ellen Dissanayake is an affiliate professor in the School of Music at the University of Washington, Seattle. Her interdisciplinary work demonstrates that the arts are integral to human nature—normal, natural, and necessary. She is the author of *What Is Art For?* (University of Washington Press, 1998), *Homo aestheticus* (translated into Chinese and Korean; University of Washington Press, 1992), and *Art and Intimacy* (University of Washington Press, 2000).

Alice Dreger is the author most recently of *Galileo's Middle Finger: Heretics, Activists, and the Search for Justice in Science* (Penguin, 2015). She is a recipient of a Guggenheim Fellowship, has been named a Visionary by UTNE Reader, and has had her work included in Norton's annual *Best Creative Nonfiction*. Her work appears in scholarly journals in medicine, the sciences, and the humanities; and she also writes for the mainstream press, including *the New York Times, The Atlantic*, and *Pacific Standard*.

Herbert Gintis is external faculty at the Santa Fe Institute. His recent books include *A Cooperative Species* (Princeton University Press, 2011), *The Bounds of Reason* (Princeton University Press, 2009), *Game Theory Evolving* (Princeton University Press, 2009), and *Moral Sentiments and Material Interests* (MIT Press, 2005). He has written many essays on market dynamics and on the unification of the behavioral sciences.

Jonathan Gottschall is a distinguished fellow at Washington & Jefferson College. He is the author of *The Rape of Troy: Evolution, Violence, and the World of Homer* (Cambridge University Press, 2008), *Literature, Science and a New Humanities* (Palgrave Macmillan, 2008); *The Storytelling Animal: How Stories Make Us Human* (Houghton Mifflin Harcourt 2012); and *The Professor in the Cage: Why Men Fight and Why We Like to Watch* (Penguin, 2015).

Henry Harpending is a professor of anthropology at the University of Utah. He has also taught at the University of New Mexico and Pennsylvania State University. He obtained his PhD in anthropology from Harvard University in 1972. His areas of interest are demography and family organization of African foragers and pastoralists, human population genetics, and human evolution, especially the origins and dispersals of modern humans in the later Pleistocene. He is co-author, with Gregory Cochran, of *The 10,000 Year Explosion: How Civilization Accelerated Human Evolution* (Basic Books, 2009).

John Hawks is an associate professor of anthropology at the University of Wisconsin, Madison. He has published widely on human evolution. His web blog is the most frequently visited anthropological site on the Internet (http://johnhawks.net/weblog).

John A. Johnson is an emeritus professor of psychology at the Pennsylvania State University, DuBois. He has published widely on the personality and evolutionary psychology of moral and educational development, career choice, and work performance. He co-edited *Advanced Methods for Conducting Online Behavioral Research* (American Psychological Association, 2010). His research on psychology and the arts has been featured by *Psychology Today, Omni, the New York Times*, and *CBS Radio*.

Daniel Kruger is an assistant research professor at the University of Michigan. He has authored and co-authored more than two dozen peer-reviewed articles. His studies have been published in academic journals focusing on evolutionary psychology, and his research has been featured in more than 100 media reports.

Dan P. McAdams is the Henry Wade Rogers professor of psychology and chair of the psychology department at Northwestern University. His research focuses on personality development across the human life course. He is the author (most recently) of *The Redemptive Self: Stories Americans Live By* (Oxford University Press, 2006), *George W. Bush and the Redemptive Dream: A Psychological Portrait* (Oxford University Press, 2011), and *The Art and Science of Personality Development* (Guilford, 2015).

Barbara Oakley is a professor of engineering at Oakland University in Michigan. Her research focuses on the complex relationship between neurocircuitry and social behavior. Oakley is a Fellow of the American Institute of Medical and Biological Engineering. Her books include *Evil Genes* (Prometheus Books, 2007), *Pathological Altruism* (Oxford University Press, 2012), *Cold-Blooded Kindness* (Prometheus Books, 2011), and *A Mind for Numbers* (Tarcher, 2014).

Massimo Pigliucci is an evolutionary biologist and philosopher of science, and the K.D. Irani professor of philosophy at the City College of New York. His research interests focus on the conceptual structure of evolutionary theory, as well as on the demarcation between science and pseudoscience. He is the co-editor (with Maarten Boudry) of *The Philosophy of Pseudoscience* (University of Chicago Press, 2013).

Michael R. Rose's 1991 book *Evolutionary Biology of Aging* (Oxford University Press) offered a view of aging that was a complete departure from the views that had dominated the aging field since 1960. In 1997, Rose was awarded the Busse Research Prize by the World Congress of Gerontology. In 2004, he published a technical summary of his work, *Methuselah Flies* (World Scientific, 2004), on the postponement of aging in *Drosophila*, followed by a 2005 popular book, *The Long Tomorrow* (Oxford University Press). His most recent book, with L.D. Mueller and C.L. Rauser, is *Does Aging Stop?* (Oxford University Press, 2011) He is currently a professor at the University of California, Irvine.

Catherine Salmon is a professor of psychology at the University of Redlands in southern California. Her research interests focus on evolutionary approaches to parental investment and sexuality, including reproductive suppression and pornography. Her books include *Warrior Lovers: Erotic Fiction, Evolution, and Female Sexuality* (co-author Don Symons; Yale University Press, 2003) and *The Secret Power of Middle Children* (co-author Katrin Schumann; Hudson Street, 2011). She is co-editor of *Evolutionary Psychology, Public Policy and Personal Decisions* (with Charles Crawford; Lawrence Erlbaum, 2004), and *Family Relationships* (Oxford University Press, 2007) and *The Oxford Handbook of Evolutionary Family Psychology* (with Todd Shackelford; Oxford University Press, 2011).

David Sloan Wilson is distinguished professor of biological sciences and anthropology at Binghamton University. His books include *Unto Others: The Evolution and Psychology of Unselfish Behavior* (with Elliot Sober; Harvard University Press, 1998); *Darwin's Cathedral: Evolution, Religion and the Nature of Society* (University of Chicago Press, 2002); *Evolution for Everyone: How Darwin's Theory Can Change the Way We Think about Our Lives* (Delacorte, 2007); and *Does Altruism Exist? Culture, Genes, and the Welfare of Others* (Yale University Press, 2014). With Jonathan Gottschall, he is the co-editor of *The Literary Animal: Evolution and the Nature of Narrative* (Northwestern University Press, 2005).

Edward O. Wilson is honorary curator in entomology and university research professor emeritus at Harvard University. He has received more than 100 awards for his research and writing, including the U.S. National Medal of Science, the Crafoord Prize, and two Pulitzer Prizes in nonfiction. He is considered one of the world's foremost biologists and naturalists.

INDEX